Music Science

RIVER PUBLISHERS SERIES IN SIGNAL, IMAGE AND SPEECH PROCESSING

Series Editors:

MONCEF GABBOUJ
Tampere University of Technology
Finland

THANOS STOURAITIS
University of Patras, Greece
and
Khalifa University, UAE

Indexing: All books published in this series are submitted to the Web of Science Book Citation Index (BkCI), to SCOPUS, to CrossRef and to Google Scholar for evaluation and indexing.

The "River Publishers Series in Signal, Image and Speech Processing" is a series of comprehensive academic and professional books which focus on all aspects of the theory and practice of signal processing. Books published in the series include research monographs, edited volumes, handbooks and textbooks. The books provide professionals, researchers, educators, and advanced students in the field with an invaluable insight into the latest research and developments.

Topics covered in the series include, but are by no means restricted to the following:

- Signal Processing Systems
- Digital Signal Processing
- Image Processing
- Signal Theory
- Stochastic Processes
- Detection and Estimation
- Pattern Recognition
- Optical Signal Processing
- Multi-dimensional Signal Processing
- Communication Signal Processing
- Biomedical Signal Processing
- Acoustic and Vibration Signal Processing
- Data Processing
- Remote Sensing
- Signal Processing Technology
- Speech Processing
- Radar Signal Processing

For a list of other books in this series, visit www.riverpublishers.com

Music Science

Marcelo Sampaio de Alencar

Institute of Advanced Studies in Communications
Federal University of Bahia (UFBA)
Brazil

LONDON AND NEW YORK

Published 2019 by River Publishers

River Publishers

Alsbjergvej 10, 9260 Gistrup, Denmark

www.riverpublishers.com

Distributed exclusively by Routledge

4 Park Square, Milton Park, Abingdon, Oxon OX14 4RN

605 Third Avenue, New York, NY 10158

First published in paperback 2024

Music Science / by Marcelo Sampaio de Alencar.

Routledge is an imprint of the Taylor & Francis Group, an informa business

Publisher's Note
The publisher has gone to great lengths to ensure the quality of this reprint but points out that some imperfections in the original copies may be apparent.

While every effort is made to provide dependable information, the publisher, authors, and editors cannot be held responsible for any errors or omissions.

ISBN: 978-87-7022-130-6 (hbk)
ISBN: 978-87-7004-340-3 (pbk)
ISBN: 978-1-003-33889-5 (ebk)

DOI: 10.1201/9781003338895

This book is dedicated to my family.

Contents

Preface

This is a book about music science, therefore, the readers are advised that the main objective of this book is to explain, more than teach, music. But, of course, the book contains music theory and more.

To fully enjoy the analysis of the embedded musical stuff, and learn to read the staff, the reader will need a certain background in mathematics, at the calculus level.

Few books have been published that cover all the subjects that are needed to really understand the very fundamental concepts of music science and music theory, including the physics and the mathematics behind it.

Some books, which deal with both subjects, are destined to very specific audiences, of expert musicians, mathematically inclined graduates, or are destined to an academic audience. The vast majority of the books introduce the subject at a very basic, or technical, level.

This book has the objective of making the subjects clear to a broader audience, by describing them in detail, by revealing relevant ideas, and by giving the reasons for each musical concept.

Brief Description of the Book

The book begins with the historical evolution of music, and how mathematics permeates that progress. An introduction to the analysis of signals in time and frequency is presented. Features and mathematical aspects of sound are discussed, including vibration and timbre aspects.

The book presents a review of existing voice models and discusses the voice production, sound perception, music characteristics and acoustics, tempo, rhythm, and harmony. Musical theory is presented, encompassing staff, notes, alterations, keys and intervals, tones, and associated frequencies.

The creation of major and minor scales is emphasized, along with a study on consonance and dissonance, measure, metric, tempo markings, and dynamics. The book also explains the chord formation, and discusses melody and composition.

The book has five appendices, including an appendix on the basic differentiation and integration theorems, another with useful Fourier tables, and an appendix featuring the notes, their frequencies, and wavelengths. The book also has a complete glossary of music terms, and a biography of the author.

This book is aimed at musicians, scientists, engineers, mathematicians, and physicists, and computer analysts. It is also useful for communication, information technology professionals and music lovers. It is expected to be used as a textbook for courses in music science or music theory.

As Guillaume de Machaut (c.1300–1377), the great French musician, poet and composer, once said *"Musique est une science qui veut qu'on rie et chante et danse."*[1]

Contents of the Chapters

The book has 17 chapters, which can be read in a non-sequential manner, if the reader has specific interests. Chapter 1 discusses the philosophy and fundamentals of music science, and its relation to other subjects. Chapter 2 presents a historical evolution of music, and discusses musical eras.

The relation between music science, mathematicians, and mathematics is presented in Chapter 3. Chapter 4 introduces the analysis of signals in time and frequency, and presents Fourier theory. Chapter 5 presents applications of Fourier transform. The properties and applications of Fourier transform form the subject of Chapter 6.

Chapter 7 presents the fundamental definitions of musical science, and musical terminology. The features of the sound production, including vibration and timbre are presented in Chapter 8. Chapter 9 discusses music characteristics, acoustics and its relation to cognition, including the effects of tempo, rhythm and harmony. Chapter 10, written by Prof. Raissa Bezerra Rocha, discusses voice models and voice production, along with sound perception.

Chapter 11 discusses the creation of major scales and the formation of minor scales, the circle of fifths, chroma, and pitch perception. The study on consonance and dissonance is the subject of Chapter 12. Measure and metric, note definition, tempo markings, and dynamics are presented in Chapter 13.

Music theory and chord formation are introduced in Chapter 14, along with a discussion on chord formation, triad, alterations, keys and intervals,

---

[1]"Music is a science that would have us laugh and sing and dance." [Yudkin, 2013].

tones, and associated frequencies. The basics of harmony, melody, and composition form Chapter 15.

Musical instruments are presented in Chapter 16, along with the usual formation of orchestras and rock bands. Chapter 17 presents the main music styles, that include classic, modern, popular, and human voice styles.

The book is completed by five appendices. Appendix A presents some basic differentiation and integration rules. Appendix B presents the properties of Fourier series and transform, and includes tables of Fourier transforms to help the reader. Appendix C has an exhaustive list of notes, including their frequencies and wavelengths. Appendix D has a comprehensive glossary of music terms and acronyms. Appendix E for the curious reader, presents the biography of the author.

Marcelo Sampaio de Alencar

Acknowledgements

The author is grateful to the colleagues and students of the Institute of Advanced Studies in Communications (Iecom), for technical contributions and useful discussions.

The author is also grateful to all the members of the Communications Research Group, certified by the National Council for Scientific and Technological Development (CNPq), at the Federal University of Campina Grande, and to the colleagues at the Federal University of Bahia, for their collaboration in many ways.

The author is indebted to his wife and children for their patience and support during the course of the preparation of this book. The author thanks Raissa Bezerra Rocha, who wrote the chapter on Voice Models, and Thiago T. Alencar, who translated selected parts of the book.

The author acknowledges the fruitful comments and suggestions given by professors Washington Araujo Neves and Jorge José Ferreira de Lima Alves, from the Federal University of Campina Grande.

Finally, the author is thankful to Mark de Jongh, from River Publishers, who strongly supported this project from the beginning, and to Rajeev Prasad and Junko Nakajima who supported the preparation of the book, and helped with the administrative, editing, and reviewing processes.

List of Figures

List of Tables

1

Elements and Philosophy of Music

"After silence, that which comes nearest to expressing the inexpressible is music."

Aldous Huxley

1.1 The Logic of Sound

Philosophy, from the Greek expression "the love of knowledge," probably coined by Pythagoras of Samos, (c.570–c.495 A.C.), the Greek philosopher and mathematician, is the study of fundamental and general problems, concerning to existence, knowledge, values, reason, the mind, and language. Figure 1.1 shows a picture of Pythagoras of Samos.

The main areas of philosophy include metaphysics, which deals with the fundamental nature of reality and being; ethics, which deals with the study and reflection about morals; epistemology, which defines the basis of knowledge, its limits and validity.

Philosophy also includes logic, which discusses the forms of argumentation; politics, which deals with organization, direction, and administration of the state; science, which deals with the systematization of knowledge; history, which studies the human being and its actions through time and space, besides the analysis of processes and events that occurred in the past; and aesthetics, the study of nature, beauty, and the foundations of art.

In the ancient classical era, philosophy was usually divided into three main branches: natural philosophy, metaphysical philosophy, and moral philosophy. The liberal arts, terminology organized in the middle ages to define the teaching methodology, included the trivium, with rhetoric, logic, and grammar, and the quadrivium, with music, arithmetic, geometry, and astronomy.

Natural philosophy, which later became known as physics, included the study of the real world. It was divided in time into several natural sciences,

1

Figure 1.1 Pythagoras of Samos. (Adapted from the Diocesan and County Library in Skara, Sweden.)

encompassing astronomy, physics, chemistry, biology, and cosmology. The practical arts included medicine and architecture.

The metaphysical philosophy is the study of existence, primary cause, logic, among other abstractions. It originated the formal sciences, such as logic, or dialectics, mathematics, philosophy of science, and epistemology.

The moral philosophy, or ethics, represents the study of what is right or wrong, justice, virtue, and goodness. It gave rise to the social sciences, which include the theory of value, with aesthetics, ethics, and political philosophy. Music is, sometimes, associated with aesthetics, which comes from the Greek word for perception, sensation or sensibility.

The study of music in the quadrivium was the classic subject of harmonics, the study of proportions among the musical intervals, created by the division of the monochord, the basic instrument used by Pythagoras to determine the fundamental proportions that guided the assembly of the scales.

In the contemporary university syllabus of arts, the quadrivium could be considered a study of the numbers and their relations with space and time. In this context, arithmetic is the pure number, geometry represents the number in space, astronomy associates the number to space and time, and music is the number in time.

In Pythagoras' time only the planets visible to the naked eye were known, like Mercury, Venus, Mars, Jupiter, and Saturn. According to the knowledge of the time, the five planets plus the Sun and the Moon, revolved around the Earth. Pythagoras postulated, then, the existence of seven notes in the musical scale, associated with the planets.

According to him, these heavenly bodies would move with speeds that had the same relation to the musical scale notes. Therefore, Saturn was associated with the note Si (B), Jupiter with Do (C), Mars with Re (D), the Sun with Mi (E), Mercury with Fa (F), Venus with Sol (G), and the Moon with La (A).

It is no coincidence that many musicians are also mathematicians, physicists, engineers, or scientists. Furthermore, many composers use number, algorithms and logic to produce their musical works, like the Greeks used to associate the movement of planets and stars to the harmony of the spheres or universal music.

1.2 The Audio and the Video

Sound can be defined as an oscillation in the pressure of the medium, like the air, that propagates as a function of internal forces, which can be, typically, elastic or viscous. However, sound can also be described as the subjective sensation caused, or stimulated, by that oscillation [Alencar, 2018f].

Human speech, in turn, is a product of the pressure exerted by the lungs, which creates a vibration in the vocal chords, or folds, and produces phonation of the glottis, the final portion of the larynx. That phonation is modified by the vocal tract and the mouth, to produce the distinct vocalic and consonantal sounds. The sound can, thus, be considered the carrier of speech information.

The human ear is sensitive to sounds in the spectrum from 20 to 20 thousand cycles per second (from 20 Hz to 20 kHz). This sensibility is superior to the spectral band of speech, which practically has no energy above 5 kHz. Probably, a result of evolution, as a precaution against predators, or natural phenomena, or to alert about prey and the occurrence of water, in the distant past of the human species.

Phonation, or voicing, is then the process by which the vocal chords produce certain sounds, from an almost periodic vibration, caused by the passage of air coming from the lungs, which provokes a set of small explosions, generating a series of cyclestationary impulses.

The passage of air through the open vocal folds is used to generate sibilant or hollow phonemes, for example, with noisy characteristics, integrating speech.

Although speech, in its genesis, is formed by discrete pulses, emitted in almost regular intervals by the vocal chords, the vocalic sounds of speech are generally perceived as continuous signals, in a process analogous to the perception of the sequence of photograms of a movie, or the frames of a video, as continuous movement.

That is a result of, in the case of video, the natural limitation of the retinas excitation capacity, associated with the brain processing time, which, in the case of the frames of a video, does not allow the perception of the individual images, if they are shown at a rate of over 10 to 15 images per second. Thus, the series of images is perceived by the brain as the continuous movement of a movie.

The auditory system has a similar limitation, and the succession of pulses generated with the abrupt opening of the vocal chords, by the air pressure produced in the lungs, ends up being perceived as a continuous audio, when it occurs at a rate over 10 to 15 pulses per second, depending on the listener. This sound, perceived as continuous, is only the result of the discrete succession of pulses, which the brain cannot keep up with.

This way, the rhythmic movement of the pulses that emerge from the glottis, with their interval marking, starts to, while the frequency of the vocal fold vibration increases, acquire a certain harmony, with the perception of pitch, or frequency of the sound. These pulses, modulated by the specific tone conferred by the vocal tract, produce the characteristic melodies of the human voice.

Melody recognition involves the perception of pitch relations. In fact, listeners are able to recognize the same melody in different registers, for instance, played on a piano or on a guitar. This means that the pattern of notes is more important than the actual key [Patel, 2010].

1.3 Made of Sound and Silence

Besides the sonorous phonemes, generated from the vibration of the vocal chords, the vocal tract also produces hollow sounds, when the glottis is totally open, allowing the passage of the air current from the lungs; in this case, the chords do not vibrate to produce the sound [Alencar, 2018c].

This free passage of the air usually leads to the production of consonantal sounds, which are modulated by the positional relation between the tongue, the lips, the teeth, the alveoli and the hard and soft palates.

However, another option for the complete opening of the vocal chords is the production of the whistle, a sound with a defined frequency, generated by the continued expiration through the mouth, which can be used, for example, to produce music.

In this case, a certain turbulence is necessary for the production of the sound, and the mouth must serve as a resonance box to reinforce the resulting sonority, acting as a Helmholtz resonator.

Helmholtz resonance is a sound-producing phenomenon which occurs when the air goes through a cavity. Hermann Ludwig Ferdinand von Helmholtz (1821–1894) was a German mathematician, doctor, pianist and physicist who demonstrated this effect, using an equipment built by himself.

Helmholtz also explained that the ghost tone, sometimes called resulting note, a low pitch tone produced when two notes sound together, is an actual physical effect [Maor, 2018]. Figure 1.2 shows a picture of Hermann Ludwig Ferdinand von Helmholtz.

The resonant cavity of the mouth produces the whistle, in a certain frequency, or pitch. It is then modulated by the position of the tongue. For instance, it is possible to produce hisses in the frequencies of 500 Hz, 1000 Hz, 2000 Hz and 4000 Hz, which are harmonic.

Usually, by the dimension of the vocal tract, men produce deeper basic whistles, with a lower frequency, in the zone of the La (A) note, which is 440 Hz. Women produce higher pitched hisses, in consonance with the dimension of their vocal tract. The hisses contain other frequencies, which form the voice timbre.

Figure 1.2   Hermann Ludwig Ferdinand von Helmholtz. (Adapted from Creative Commons – https://creativecommons.org)

Several instruments use resonance to produce sound, such as, for instance, the flute and the tuba. A resonant tube produces a specific frequency, according to its dimension. Other instruments use a cavity to amplify the sound that is produced, like the violin or the guitar.

Curiously, the vocal tract, since it is flexible, has the capacity to produce sounds in a broad, band or *tessitura*, which would be difficult for fixed dimension, or invariant in time, instruments.

As an example, the tuba has a large dimension, so it can produce deep or low sounds. The tuba with the deepest sound is the contrabass, which produces fundamental sound in the band of 32 Hz, for tuning in Do (C). The main tube of that tuba can reach up to 4.9 m long. For reference, the deepest note in the known repertoire can reach 16 Hz. This frequency is inaudible, but it can be felt by the sense of touch.

The tactile sense is composed of a very fine network of receptors in the skin, that forms the body's largest sensory system. The receptors sense pressure on the skin, and that is how it is possible to feel touch. Because there are many sensory nerves, the lightest touch is felt. Therefore, although the hearing sense does not perceive the sound at a frequency below 20 Hz, the body can feel it through the skin.

The band of frequencies from 10 Hz to 15 Hz, which cannot be reached by the deepest tuba, usually represents an inflexion region, which separates the duration of the sound, a typical characteristic of the marking obtained with percussion; of its pitch, or perception of frequency, which is an aspect related to the harmony in musical theory [Wisnik, 2017].

1.4 The Quantum Nature of Music

When the guitar string is strung by a player, a sound in a specific frequency is produced, which the non-linear nature of the instruments' constituents decimates into harmonics, generating the characteristic guitar timbre [Alencar, 2018a].

The main frequency, when perceived by the auditory system, generates a sonority, producing the perception of a tone, with a certain tuning. Certain frequencies were chosen to compose the musical scales, through time, in detriment of others, and that choice quantizes, or quantifies, the music.

In the case of the diatonic scale, the La (A) scale was agreed upon to correspond to the frequency of 440 Hz, which can be found above, or to the right of the central Do (C) scale of the piano, whose frequency is 261.6 Hz, in the Sol (G) key range in the musical score.

This standard was adopted by the International Organization for Standardization (ISO) as the norm ISO 16:1975, confirmed in 2017, and serves as reference for the tuning of the pitch height, or frequency, of the instruments.

This mechanical vibration could be defined as a sound quantum, or phonon, if that measure of vibrational energy had not been already defined as the unit of oscillation of the atoms of a crystal.

This concept was introduced, in 1932 by the Soviet physicist Igor Yevgenyevich Tamm (1895–1971). Tamm was awarded the Nobel prize for the discovery, in 1934, of the Tcherenkov radiation, which appears when a charged particle passes through a dielectric medium at a speed superior to the phase speed of light in that medium.

Although it might appear strange, for those who know something about the theory of relativity, produced by the most celebrated German theoretical

Figure 1.3   Albert Einstein. (Adapted from MLA style: Albert Einstein – Biographical. NobelPrize.org. Nobel Media AB 2019. Wed. 21 Aug 2019. https://www.nobelprize.org/ prizes/physics/1921/einstein/biographical)

physicist of the 20th century, Albert Einstein (1879–1955), it is possible for a particle to travel at a speed superior to that of light, as long as it is through a material medium. In vacuum, light is the limit for speed. Figure 1.3 shows a photo of Albert Einstein.

The main observable characteristic of the electromagnetic radiation produced by that superliminar particle is the intense blueish brightness emitted by an underwater nuclear reactor.

It is curious to know that the Tcherenkov radiation had been theoretically foreseen by the British scientist Oliver Heaviside (1850–1925), in a series

Figure 1.4   A portrait of Oliver Heaviside. (Public domain. Wikimedia Commons. Original from Smithsonian Libraries.)

of papers published between 1888 and 1889. The radiation was named after the Russian physicist and Nobel prize winner Pavel Alexeevitch Tcherenkov (1904–1990), who discovered and interpreted the effect.

Heaviside was very versatile and, from a profound knowledge of electricity, used the complex number theory to study electric circuits, and later on reformulated the field equations established by James Clerk Maxwell (1831–1879).

Maxwell' formulation consisted of 20 complicated equations, with 20 unknowns. Heaviside created the modern vectorial notation, and reduced 12 of Maxwell's equations to the well known four vectorial equations. Figure 1.4 shows a portrait of Oliver Heaviside.

He also formulated the vectorial analysis, invented the unit step function, known as Heaviside's function, to compute the current when the circuit is turned on. He was the first to use the impulse function, also known as Dirac's delta, which is fundamental to model singularities, in applications that vary from probability distributions to black holes.

Dirac's delta, the impulse generalized function, is named after Paul Adrien Maurice Dirac (1902–1984), an English theoretical physicist, who made fundamental contributions to the early development of quantum mechanics and quantum electrodynamics.

Heaviside suggested the existence of gravitational waves, later also studied by Einstein, that could reveal the hidden secrets of the universe, such as, the existence of black energy and matter, or the strange worm holes between multiverses.

And, if it was not enough, Heaviside developed the transmission line theory, and created a technique to solve linear differential equations, using algebraic equations, called operational calculus, which is equivalent to the well known Laplace transform.

He published several scientific articles, in prestigious journals, received the Faraday medal, named after the English scientist Michael Faraday (1791–1867), and became a member of the Royal Society, even though he did not have a college degree, and lived most of his life unemployed and in seclusion.

The audiences of the famous musical Cats, composed by Andrew Lloyd Webber (1948–) based on Thomas Stearns Eliot's (1888–1965) book "Old Possum's Book of Practical Cats," usually do not realize that, at the climax of the piece, when the choir members sing "Up, up, up to Heaviside layer," they are referring to the ionosphere, the ionized layer used to transmit short radio waves, which, by the way, was also idealized by Oliver Heaviside.

1.5 The Sound Perception

The long wavelength phonons, such as those produced by the piano's strings, convey sound, the mechanical oscillation that propagates through the air, and can be perceived by the auditory system. The word comes from the Greek *phone* that also generated the words phoneme and telephone [Alencar, 2018b].

The continuous sound range has been quantized in the musical scales, used in diverse cultures. The seven-note sequence, Do, Re, Mi, Fa, Sol, La, and Si, represents the seven quanta of music in the diatonic scale, generally used in the Western part of the world. But, the diatonic scale could be very old. A Sumerian clay signboard, dating circa 1400 A.C, displays a love song that uses a scale similar to the diatonic.

Aristoxenus of Tarentum (c.375–300 BC), a Greek philosopher of Pythagorean tradition, and a pupil of Aristotle, wrote a musical treatise, entitled

Elements of Harmony, in which he exposed the Greek musical system, based on the perfect fourth interval [Ball, 2010].

The Eastern part has an affinity for the pentatonic scale, which is equivalent to the five black notes of the piano keyboard, but there is an outstanding variety of scales in use in the world, some with dozens of musical notes.

For instance, the current Indian music system is based on two important definitions. The modes of the classic Indian music are called *raga*, the melodic form, and *taal*, the rhythmic aspect. The octave spans 22 equal divisions, called *sruti*, and the tiny fractions of Indian music (microtones) are called *shrutis*, from the Sanskrit verb to hear.

The inclusion of flats (*bemols*) or sharps (*diesis*) gave birth to the chromatic scale, with 12 notes. The other intermediary frequencies are usually discarded from the typical musical scale, but remain hidden in the instrument *timbre*.

In the Western world, the mathematician and philosopher Pythagoras, more frequently associated to the theorem that bears his name, established in the 6th century BC the first numerical relation associated with the musical harmonic series.

He used the monochord, a single string instrument, to demonstrate the scale formation from the division of the string into defined fractions, and used weights to produce the adequate tensions to generate the diverse musical tones.

As he varied the monochord string length and applied tension, he established the intervals between the musical notes, and defined in this way the harmonic series that is the basis for the scales in the Western world, the Greek modes, seven different models for the major natural scale.

The harmonic series is, thus, the sound resulting from the mechanical vibration of an object, that produces a fundamental frequency, aside from other multiple frequencies. The composition of the differences between the frequencies form the sound pitch intervals and establishes the musical scale.

Pythagoras imagined the music as an exact science, such as mathematics, related to the transcendental images of utopia, analogous to the Hinduist theory, but also as a cosmological science, related to the harmony of the spheres, the music associated to the seven celestial errand bodies, by this reason called planets, that were perceived at that time.

The quadrivium, the second part of the syllabus established by Plato in his book *The Republic*, and that lasted for nine centuries, included music as one of the compulsory subjects, along with astronomy, arithmetic, and geography. The preparatory work to study philosophy or theology.

The quadrivium was preceded by the trivium, composed of grammar, rhetoric, and logic, the, so-called, trivial disciplines. The study of the seven liberal arts, along the centuries, included the trivium and the quadrivium. For many centuries, and nowadays, the composers use mathematics to produce, or inspire, their musics, a return to the Hellenistic genesis of the sound.

It is interesting to note that the study of music began with a single string instrument, such as the simple and famous Brazilian *berimbau*, which is the basis for the mix of dance and martial art known as *capoeira*, that define both its movements and its rhythm. The rhythm is related to periodicity, as a systematic mosaic of sounds, and involves timing, accent, and grouping [Patel, 2010].

To play the *berimbau*, interestingly, it is necessary to master seven components, the drumstick (*baqueta*), the doubloon (*dobrão*), the gourd (*cabaça*), the lintel (*verga*), the rope (*corda*), the lashing of the head (*amarração da cabeça*), and the *caxixi*.

The *berimbau* was a theme of one of the most emblematic songs composed by Baden Powell and Vinícius de Moraes. This song originated the series of afro-sambas, that created its own school of musical composition, and enriched the Brazilian popular songbook.

1.6 The Material and the Virtual

Since immemorial times, the world of arts, different from the technology environment, always understood well the difference between two important concepts: the material and the virtual, which represent the substrate and the creation [Alencar, 2018d].

In music, for instance, a good artist can be a *virtuoso*, even when playing a cheap guitar, an instrument of a lesser quality, that could worth only a fraction of the legal rights of a famous song that is listed in a music recital.

In communications, for example, in the beginning of telephony, the equipment was basically hardware. The fixed telephone did not carry any software. In truth, it did not have enough memory to store the telephone numbers.

All the necessary software, including the dialing procedure and the memory for the digits, had to be stored in the user's mind or, maybe, in the operator's mind. Some virtuous operators, very regarded and requested at that time, could memorize all the telephone numbers of a city of the size of Rio de Janeiro.

The cell phone, when it was launched, in 1979, was a bit more than a transceiver, a combination of transmitter and receiver, with a keyboard, aside from the electronic circuitry, the microphone, and the headphone. It had some memory to store the telephone numbers, and counted on a limited processor, that run a basic operator system.

As time went by, the cell phone added new functions and facilities, including a photographic camera, and later on, a video camera. Then came along an electronic agenda, a calculator, and an increased screen, that ended up occupying the whole device. The processor became a full computer, more powerful than a mainframe computer from the past decades.

The number and variety of available programs is huge, nowadays, for any type of application, and new ones appear all the time, including fanciful music apps, to help find, play, store, elaborate, produce, and market songs.

In fact, a communication system has more similarities with the music environment than one can, at first, imagine. For instance, the transmission frequency is akin to the music pitch, the harmonics can be compared to the music partials, and the channels would be the different instruments in polyphonic music.

The list can go on, because the channel power relates to the music loudness, the pulse is the equivalent of the beat in music, and the communication data rate is related to the *tempo* in music. Finally, the spectrum occupation is quite similar to the placement of musical notes on the staff, and an orchestra communicates musical information in an organized and beautiful manner.

The communication hardware changes became so frequent, that demanded the creation of the software defined radio concept, that is, even the basic functions of the device, that were in the past executed by the hardware, such as modulation and audio processing, are now executed by software routines, that run on standard platforms.

This process, called virtualization, also reach base stations, that control the incoming calls to the access points of the cellular system. The telephone exchanges, responsible for switching and routing, are now virtualized. Interestingly, they used to be called stored program control exchanges.

In the end, the virtualization, that transforms hardware into software, will reach the whole network, the optical fibers, the radio links, and the set of satellites that orbit planet Earth. In a sense, the poetic image created by Pythagoras, that related the movement of the celestial bodies to music, is not very far from becoming true.

Because life imitates art, the communications engineers tried to go in search of lost time. Slowly, the physical layer, the material substrate or hardware, gives way to the virtual, the immaterial, the software.

As Albert Einstein once said, "After a certain high level of technical skill is achieved, science and art tend to coalesce in aesthetics, plasticity, and form. The greatest scientists are always artists as well."

1.7 The Organized Sound

Edgard Victor Achille Charles Varèse (1883–1965), a French engineer and composer, a pioneer of electronic music, at the very beginning of electronics, became well known for having abandoned the traditional composition method in favor of using the proper sound material. Figure 1.5 shows a portrait of Edgard Varèse.

Figure 1.5   A portrait of Edgard Varèse. (Public domain. Wikimedia Commons.)

To break with the past, he destroyed all the originals of his own previous compositions, because, to him, they followed conservative patterns. One of his fundamental ideas was the notion of music as organized sound [Alencar, 2018e].

Pythagoras of Samos, well known because of the theorem that bears his name, made the initial tentative to organize the music, when he established the first rules of sound consonance. Pythagoras perceived the harmonious relation between the sound in a certain frequency and double this frequency, the octave. He also noted that certain ratios of integer numbers, when applied to a string length, produced harmony.

For example, when a string is divided into three parts and held by one of the marked points, it gives the ratio 2/3 for the larger resulting part. If the longer portion of the string is plucked, a sound is produced which harmonizes with the open string. The produced note, that has a frequency that is 50% higher than the original one, that is known as the tonic, is called the fifth, or dominant.

For the acoustic guitar, the sixth string, the thicker one, is known as the Mi (E) string. If this string is held by the seventh fret, which permits the ratio 2/3 mentioned, and plucked, it produces the note Si (B), the fifth note of the scale that begins with the tonic Mi (E). Besides, the 12th fret indicates half the string, and produces the same Mi (E) note, but with a frequency that is double that produced by the open string.

The ratio 3/4, on the other hand, produces a frequency that is lower than that produced by the fifth note, and higher than the original note, exactly by the fraction 4/3. It is called the fourth, or subdominant, and also maintains some harmony with respect to the tonic of the scale.

It is not a surprise that the note Fa (F) appears so often in songs composed in the key of Do (C), considering that it is the fourth of the Do scale. By the way, the note Sol (G) also appears very often.

For the scale that has Do (C) as tonic, the remaining notes follow the tonal degrees of the diatonic scale, with the defined integer ratios, in relation to the tonic: Re (D) is the supertonic (9/8), Mi (E) is the mediant (81/64), Fa (F) is the subdominant (4/3), Sol (G) is the dominant (3/2), La (A) is the superdominant or submediant (27/16), and Si (B) is the subtonic or leading tone (243/128).

By the way, the frequency of the central Do (C) of a piano keyboard is 261.6 Hz (cycles per second). The note Do′ (C′), in which the accent mark indicates that the note is one octave above the central Do (C), has a frequency of 523.2 Hz, that corresponds to the lowest Do (C) of the acoustic guitar, that is obtained by pressing the La (A) string on the third fret.

There is no mystery in obtaining the previous ratios. For example, for the note Re (D), the supertonic of the Do (C) scale, it is only necessary to remember that the perfect fifth of Do (C), obtained by the multiplication of its frequency by 3/2, is the note Sol (G).

On the other hand, the perfect fifth of Sol (G), obtained by the same procedure, is Re (D), one octave above, that is, with a frequency that is double the original one. Therefore, just do $3/2 \times 3/2 = 9/4$, and divide the result by two, to obtain 9/8, that is the correct fraction for the note Re (D).

The music, as seen, is really the organized sound, as Edgard Varèse once said. The nature sounds are produced by several animals, such as the birds, the cetaceans, the chimpanzees, the bats, but only with the humans the sound became organized, creative and mathematical, to the point of producing music. But, the reason for the appearance of music among the humans continues to be a mystery.

1.8 Music and Poetry

Not everyone likes poetry, but it is difficult to find someone that does not have a favourite song. This is quite curious, because a song is only poetry with the addition of music. That is, a song is just a chanted poetry.

What differentiates a song from recited poetry are the variations in parameters, such as, frequency, intensity, rhythm and duration of the voiced sounds, that occur when one sings. For instance, Figure 1.6 illustrates the relation between some basic music parameters, for *bossa nova* and hip hop.

Poetry is a literary gender characterized by the composition of structured verses, in harmonious form. It is a manifestation of beauty and aesthetics, that is pictured by the poet in words.

In a figurative meaning, poetry is everything that moves, that sensitizes and awakens feelings. It is any form of art that inspires and enchants, that is sublime and beautiful.

There are certain formal elements that characterize the poetic text, such as, for instance, the rhythm, the verses, and the stanzas, that define the poetry metric. The use of specific literary resources is what distinguishes the style of a poet.

For example, the free verses do not follow a metric. The author has complete freedom to define a rhythm and create a standard of its own. This type of poetry is also called modern poetry.

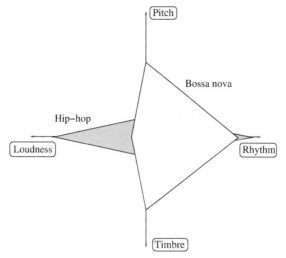

Figure 1.6 The relation between some basic music parameters, for *bossa* nova and hip hop.

Poetry, or lyric text, is one of the seven traditional arts, by which the human language is used with aesthetic or critic purposes, that is, it portrays a scene in which anything can happen, depending on the imagination of the author and the reader.

It is difficult to know how exactly the music of the ancient Greece would sound, but the Greek music was probably of vocal tradition, derived from verses that were sung accompanied by an instrument such as the lyre, from which the word lyrics came in the first place. The kithara and the flute were also used in those occasions [Ball, 2010].

A song is a relatively short composition, that combines a certain melody, a music, with a literary text, the lyrics. Of course, songs can also be a musical composition without lyrics, that are called instrumental songs.

When the song is not accompanied by any musical instrument, or by a recording of instrumental sounds, it is called *a cappella* music. In classic music, the word melody designates a musical part, the Italian word *aria*, that comes from an opera plot, carries the meanings of love or tragedy, and the German word *Lied*, that describes the setting of poetry to classical music, to create a piece of polyphonic music. The lyrics is typically of a poetic or rhythmic nature.

According to Luiz Tatit (1951–), a Brazilian musician and linguist, song is lyrics and melody. When one of these factors is suppressed, there is no song. Tatit cites the Brazilian thinker, journalist and poet José Lino Grünewald

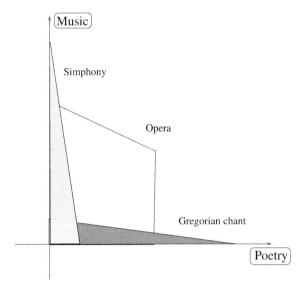

Figure 1.7   Pictorial relation between poetry and music.

Figure 1.8   Relation between poetry and music.

(1931–2000): "There is, then, a peculiar fact: verses, that when read or declaimed are simple, gain new effects in accordance with the musical track, and the way they are sung" [Tatit, 1986].

The dictionaries usually define poetry as a composition in verses, typically with a harmonious association between words, rhythms and images. This is very close to the concept of a song. The poetry and the music have always been related. Figure 1.7 illustrates the pictorial relation between poetry and music.

In Ancient Greece, a poetic composition, made to be sung, either by a voice or by choir, was called lyric, because it was accompanied by a lyre. The term lyric became synonymous of expressing feelings or intimate thoughts. Figure 1.8 illustrates the relation between a few poetry characteristics and some music features.

The song is a very old musical form. For Aristotle, there was not a marked difference between poetry and singing, because the reproduction of the former used to be done by the latter. Until the final centuries of the Middle Ages, singing and reciting poetry did not mean very different things.

2

History and Evolution of Music

"A painter paints pictures on canvas. But musicians paint their pictures on silence."

Leopold Stokowski

2.1 Music of the Evolution

How music entered the human culture remain a mystery for researchers in the area, There are no evidences that point to any evolutionary advantage that the musical practice or creation could give to be appropriated by the human genome.

However, the relation between the human being and the music is a lasting one, actually pre-historic. There are records of a bone flute, with five regularly disposed holes, to produce specific sounds, dating from 35 thousand years ago. The instrument has been found in the Hohle Fels cave, in the German Baviera, near the city of Ulm, where Albert Einstein was born [Maor, 2018].

Rhythms and sounds are part of the nature of the universe, and the primitive music was first of all rhythmic, sung and danced, before being instrumental and melodic [Sallet, 1974].

The sound production is common among birds and other animals, such as the chimpanzees, the dolphins and the bats, to cite a few, but the musical creation, as far as it is known, only happens among humans.

There is some speculation that the music could have appeared as a form of sexual attraction, considering that various species of birds, usually, males, produce elaborate sounds to lure females. This type of behavior, it seems, is more common among the men, than in women, as revealed by several published opinion polls.

In fact, a closer look at the biology of animals that can sing, mainly birds and whales, reveals that songs are typically produced by males, who use them to attract mates, or to put away competitors. In other words, the production of songs is a biologically mediated reproductive behavior [Patel, 2010].

An analysis, conducted in 2018 by Jennifer W. Shewmaker, Andrew P. Smiler and Brittany Hearon, examined sexual stereotypes in the pop music, using the Billboard Top 100 publication, between 1960 and 2008. Most of the songs were created by men, in all the genres, but in different rates.

Rhythm and Blues songs showed the lowest discrepancy (52.0%), while Rap displayed the highest imbalance between the genders (93.2%), Rock also showed a high percentage of men (77.8%), and other types of music followed the same high pattern of male influence (70.9%).

According to a recent study, published by *The New York Times*, conducted by Stacy L. Smith, from the University of Southern California, considering the 600 most successful songs, in the period from 2012 to 2017, defined by the Billboard Hot 100, with 1,239 artists, only 22.4% of them were women. The number is even lower for composers, considering that only 12.3% of them are women.

Charles Darwin (1809–1882), in his classical book "On the Origin of Species by Means of Natural Selection, or the Preservation of Favoured Races in the Struggle for Life", published in 1859, speculated on the human ability to create music. According to Darwin, music seemed not to aggregate any direct utility to the daily lives of the *homo sapiens*' ancestors. Figure 2.1 shows a picture of Charles Darwin.

Darwin argumented, in 1871, that human music would be, in fact, a sexually selected behavioral trait. Indeed, the human being would not gain a reproductive advantage thanks to a longer life, for example, but thanks to a successful courtship.

He thought that music, as an evolutionary behavior, did not present an adaptive value. The music had no relation at all to natural selection, but could present a relation to sexual selection.

Music would denote abilities that would make the musician more attractive to the opposite sex, such as coordination, determination, a good ear, and resilience, characteristics that a female would want for her off springs. As it is well known, musicians are popular among women since, at least, 35 thousand years.

2.2 The Pre-Historic Antiquity

Music was considered by the primitive peoples as having a magical origin. It used to be overall rhythmic, to be sung or danced. The first instruments of

Figure 2.1 Charles Darwin. (Adapted from https://snl.no/Charles_Darwin. Public domain.)

the pre-historic and antiquity ages were the ancestors of the current musical instruments, and included the drum, made of animal skin; the flutes, made of bones or plants; and the gongs and cymbals, made of metal [Sallet, 1976].

The Sumerians formed one of the more antique and more evolved civilizations, and produced several musical instruments, mainly used for royal and religious functions. The harp and the lyre, both stringed, were the most used instruments. The Sumerian musicians had a marked influence on the other people of the time, including the Egyptian civilization [Sallet, 1974].

The Greek term *musiké*, which means "the art of the muses" was the origin of the word music. In Greek mythology, the muses were the goddesses that inspired the poesy, the astronomy, the dance and the music.

The Pythagorean school searched for the harmony of the universe, and considered the natural numbers, and their relations, the manner to express that harmony. Pythagoras of Samos (c.570–c.497 BC), was a Greek philosopher and mathematician [Henrique, 2014].

The Pythagoreans considered that the movement of the planets, errands in a free translation, produced harmonic vibrations, which were called "music of the spheres", imperceptible to the human beings [Arbonés and Milrud, 2012].

It is interesting that this ancient concept is not too far from the theory on gravitational waves, proposed by Albert Einstein, 2400 years later, and only recently proved right by the scientists and discovered by the astronomers [Einstein, 1916].

2.3 Music Eras

It is possible to divide the history of music into distinct eras, that are identified by the peculiar style of the period [Bennet, 1986].

2.3.1 Middle Ages

The Middle Ages, or medieval period, was as an era dominated by catholic sacred music (c.450–1450), which began as a simple chant but grew in complexity in the 13th to 15th centuries by experiments in harmony and rhythm.

The composer Perotinus Magnus, also known as Pérotin, (1160–1230) and the poet and composer Guillaume de Machaut (1300–1377) are representatives of that era.

During this era, the Latin hymn in honor of John the Baptist, *Ut queant laxis*, was written, probably by Paulus Diaconus (c.720–799), an 8th-century Lombard historian. The Benedictine monk Guido d'Arezzo (c.991–c.1050), an Italian music theorist, possibly composed the melody, but it is possible that he used an existing melody. He also invented the modern musical notation.

The hymn, which belongs to the tradition of Gregorian chant, that is similar to the Greek diatonic music, is known for its part in the history of musical notation, particularly regarding solmization [de Mattos Priolli, 2013b]. The first stanza of the hymn is [Wikipedia contributors, 2019]:

Ut queant laxis
resonare fibris

Mira gestorum
famuli tuorum,
Solve polluti
labii reatum,
Sancte Iohannes.

An usual translation is: "So that your servants may, with loosened voices, resound the wonders of your deeds, clean the guilt from our stained lips, O Saint John." The hymn first syllables are used to denote the musical scale *Do, Re, Mi, Fa, Sol, La,* and *Si.* The first note *Ut* is still used in the French musical notation.

Cecile Gertken (1902–2001) wrote a paraphrase of the hymn that preserves the key syllables and somehow evokes the original meter:

Do let our voices
resonate most purely,
miracles telling,
far greater than many;
so let our tongues be
lavish in your praises,
Saint John the Baptist.

The *Organum* is a type of early French medieval polyphony dating from c.1000–1200, featuring a slow non-metered chant in the lowest voice with one or more faster metrical voices sung above, in melismatic style. It also means many notes sung on each syllable of text.

2.3.2 Renaissance

The Renaissance was an era (c.1450–1600) that witnessed the rebirth of learning and exploration. This was reflected musically in a more personal style than seen in the Middle Ages.

The important composers were Josquin des Prez (c.1450–1521), French, Giovanni Pierluigi da Palestrina (1525–1594), Italian, and Thomas Weelkes (1576–1623), English. The tonal era appeared during the end of the Renaissance, as a perfectioning of the modal practice from the Middle Ages [Almada, 2009].

2.3.3 Baroque

The Baroque era was a musical period (c.1600–1750) of extremely ornate and elaborate approaches to the arts. This era saw the rise of instrumental music, the invention of the modern violin family and the creation of the first orchestras, by Antonio Lucio Vivaldi (1678–1741), an Italian composer, violonist and priest; George Frideric Handel (1685–1759), a German, later British, composer; and Johann Christian Bach (1685–1750), a German composer and musician.

In that time, the *Collegium Musicum* was a university ensemble dedicated to the performance of early music, and the Doctrine of Affections was the baroque methodology for evoking a specific emotion using music and text.

An *episode* was an intermediary, contrasting, section of a Baroque fugue or Classic rondo form equal temperament. Now, it is a standard modern tuning system in which the octave is divided into 12 equal half-steps.

The *basso continuo* was the back up ensemble of the Baroque era, usually comprised of a keyboard instrument, a harpsichord or an organ, and a melodic stringed bass instrument, a *viol' da gamba* or *cello*.

2.3.4 Classic

The Classic era was a politically turbulent period (c.1750–1820) focused on structural unity, clarity, and balance, that featured great musicians, such as, Franz Joseph Haydn (1732–1809), an Austrian composer; Wolfgang Amadeus Mozart (1756–1791), an Austrian composer; and Ludwig van Beethoven (1770–1827), a German composer and pianist. Figure 2.2 shows a picture of Ludwig van Beethoven.

The opening section of a classic sonata form, in which the two opposing key centers are exposed to the listener for the first time, was called an exposition. It was also the opening section of a fugue.

2.3.5 Romantic

The Romantic era (c.1820–1890) saw the emergence of flamboyance, nationalism, the rise of great performers, and large concerts, aimed at middle-class paying audiences. Nationalism is a musical style that include folk songs, dances, legends, language, or other national imagery relating to a composer's native country.

Orchestral, theatrical, and soloistic music grew to spectacular heights of personal expression, and the main musicians were Wilhelm Richard Wagner

Figure 2.2 Ludwig van Beethoven. (Adapted from Creative Commons – https://creativecommons.org)

(1813–1883), a German composer, theater director, and conductor; Johannes Brahms (1833–1897), a German composer, pianist, and conductor; Franz Peter Schubert (1797–1828), an Austrian composer; Louis-Hector Berlioz (1803–1869), a French composer; Frédéric François Chopin (1810–1849), a Polish composer and pianist; and Pyotr Ilyich Tchaikovsky (1840–1893), a Russian composer.

Musikdrama is a type of ultra-dramatic German operatic theatre developed by Richard Wagner in the mid-to-late Romantic era. Wagner also introduced the overture, a one-movement orchestral introduction to an opera. Bizet and other composers, after 1850, used the term *prelude* instead to show dramatic unity between the overture and the theatrical drama that follows it.

2.3.6 Modern Era

The Modern era is a musical period, from *circa* 1890 to present, impacted by experimentation, advances in musical technology, and popular or non-Western influences.

The main musicians were Claude Debussy (1862–1918), a French composer; Arnold Schönberg (1874–1951), an Austrian, and later American, composer, music theorist, teacher, writer, and painter; Igor Fyodorovich Stravinsky (1882–1971), a Russian composer, pianist, and conductor; Aaron Copland (1900–1990), an American composer, composition teacher, writer, and conductor; and John Milton Cage Jr. (1912–1992), an American composer, music theorist, artist, and philosopher.

Musique concréte is comprised of natural sounds that are recorded or manipulated electronically, or via magnetic tape. This compositional approach was promoted by Edgard Victor Achille Charles Varèse (1883–1965), a French-born composer, in the 1950s.

2.3.7 Neoclassicism

The Neoclassicism was an early 20th Century compositional style, in which Classic forms and the aesthetics of balance, clarity, and structural unity are combined with modern approaches to harmony, rhythm, and tone color.

The movement was a reaction the chromaticism of late-Romanticism and Impressionism, and appeared in parallel with the musical Modernism, which wanted to abandon the key tonality completely. Sergei Sergeyevich Prokofiev (1891–1953), a Russian composer, pianist and conductor, and Stravinsky are well-known composers in this style.

3

Music and Mathematics

"Music is the shorthand of emotion."

Leo Tolstoy

3.1 Music in Theory

That music theory is related to mathematics is not strange to mathematicians and musicians, as several books have been published in this context. Mathematics, as discussed, defined the musical scales [da Cunha, 2008], and also gave music a geometrical and visual interpretation [Ashton, 2001].

Actually, the sequence of tones, or notes, form a finite field, or Galois field, after Évariste Galois (1811–1832), a French mathematician and political activist. A field is a fundamental algebraic structure, which is used in algebra, number theory, and many other areas of mathematics. It is a set on which addition, subtraction, multiplication, and division are defined and satisfy certain basic rules. Figure 3.1 shows a portrait of Évariste Galois, at about age 15, made by his brother.

For instance, the chromatic scale is a finite field of order 12, just like the set of hours of the day. When the clock strikes 12 hours, it moves back to the beginning and starts over again. It is the same with the musical scale. For example, the Do (C) major scale starts on Do, the first degree, and goes up to Si (B), the 12th degree, and then returns to Do again, as shown in Figure 3.2. The pentatonic and the diatonic scales are also finite fields.

Music is linked to isometric transformations, a shape-preserving transformation, or movement, in the plane or in space, such as reflection, rotation, and translation and their combinations, to harmonic symmetries and matrices. In music it is possible to do horizontal scaling, which implies a change in *tempo*, and vertical scaling, a change in the interval distance between the notes.

It is also possible to represent a composition in terms of a numerical series and, in this case, the operations are performed using modular arithmetic, for

29

Figure 3.1   Portrait of Évariste Galois (This work is in the public domain. Creative Commons).

Figure 3.2   The chromatic major scale (Adapted from The Online Piano Studio (Epianostudio), Creative Commons).

example $3 = 15 \bmod 12$, which means that 3 is the remainder of the division of 15 by 12 [Arbonés and Milrud, 2012].

3.1.1 Equal Temperament

Table 3.1, adapted from Appendix C, displays the chromatic major equal-tempered scale of Figure 3.2, beginning with the middle Do (C4), with the

Table 3.1 Sequence of notes, frequencies and wavelengths for the major equal-tempered scale

Numbers	Notes	Standards	Frequencies (Hz)	Wavelengths (cm)
n_1	Do	C4	261.63	131.87
n_2	Do#	C#4/Db4	277.18	124.47
n_3	Re	D4	293.66	117.48
n_4	Re#	D#4/Eb4	311.13	110.89
n_5	Mi	E4	329.63	104.66
n_6	Fa	F4	349.23	98.79
n_7	Fa#	F#4/Gb4	369.99	93.24
n_8	Sol	G4	392.00	88.01
n_9	Sol#	G#4/Ab4	415.30	83.07
n_{10}	La	A4	440.00	78.41
n_{11}	La#	A#4/Bb4	466.16	74.01
n_{12}	Si	B4	493.88	69.85
n_{13}	Do	C5	523.25	65.94

sequence number, the associated note, the standard notation, the frequency and the wavelength, for each note.

The equal-tempered scale was first proposed to the public, in 1584, by Chu Tsai-Yü, prince of the Chinese Ming dynasty and a musician.

The Flemish mathematician and engineer Simon Stevin (1548–1620), an associate of prince Maurice of Orange, who commanded the Dutch West India Company (WIC) in Brazil, published the same idea a year later. [Ball, 2010].

For the equal-tempered scale it is possible to compute the frequency f_n within the first octave, given the reference frequency f_0, using the formula

$$f_n = f_0 \cdot 2^{n/12}. \tag{3.1}$$

Example: The pitch of two semitones above the note La (A), whose frequency is 440 Hz, is given by

$$f_2 = 440 \cdot 2^{2/12} = 493.88 \text{ Hz}.$$

3.1.2 Just Intonation

Equal temperament is a tuning system, an alternative to the just intonation, in which the octave is divided into 12 semitones, that increase in frequency by a factor of $2^{1/12}$ or $\sqrt[12]{2} = 1.05946309436$. Therefore, the semitones do not have equal size, but, in fact, they increase in size, by the same factor, as

Table 3.2   Sizes of various just intervals compared against their equal-tempered scale

Name	Ratios	Just intervals
Unison (C)	1	1/1 = 1
Minor second (C♯/D♭)	1.059463	16/15 = 1.06666
Major second (D)	1.122462	9/8 = 1.125
Minor third (D♯/E♭)	1.189207	6/5 = 1.2
Major third (E)	1.259921	5/4 = 1.25
Perfect fourth (F)	1.334840	4/3 = 1.33333
Tritone (F♯/G♭)	1.414214	7/5 = 1.4
Perfect fifth (G)	1.498307	3/2 = 1.5
Minor sixth (G♯/A♭)	1.587401	8/5 = 1.6
Major sixth (A)	1.681793	5/3 = 1.66666
Minor seventh (A♯/B♭)	1.781797	16/9 = 1.77777
Major seventh (B)	1.887749	15/8 = 1.875
Octave (C)	2	2/1 = 2

Table 3.3   Comparison between the frequencies of the equal temperament scale and the just intervals

Notes	Standards	Tempered (Hz)	Just (Hz)	Differences (Hz)
Do	C4	261.63	261.63	0
Do#	C#4/Db4	277.18	272.54	+4.64
Re	D4	293.66	294.33	−0.67
Re#	D#4/Eb4	311.13	313.96	−2.84
Mi	E4	329.63	327.03	+2.60
Fa	F4	349.23	348.83	+0.40
Fa#	F#4/Gb4	369.99	367.92	+2.07
Sol	G4	392.00	392.44	−0.44
Sol#	G#4/Ab4	415.30	418.60	−3.30
La	A4	440.00	436.05	+3.94
La#	A#4/Bb4	466.16	470.93	−4.77
Si	B4	493.88	490.55	+3.33
Do	C5	523.25	523.25	0

the frequency gets higher. The frequency of the octave, or 12th note, is, as expected, double the frequency of the original tone, because $\left(\sqrt[12]{2}\right)^{12} = 2$.

The sizes of various just intervals compared against their equal-tempered counterparts, for the key of Do (C), are shown in Table 3.2.

A comparison between the frequencies of the equal temperament scale and the just intervals, for the key of Do (C), is shown in Table 3.3.

The chromatic sequence can be represented as

$$N = (n_1, n_2, n_3, n_4, n_5, n_6, n_7, n_8, n_9, n_{10}, n_{11}, n_{12}). \tag{3.2}$$

Table 3.4   Transposed sequence of notes, frequencies and wavelengths for the major scale

Numbers	Notes	Standards	Frequencies (Hz)	Wavelengths (cm)
n_1	Re	D4	293.66	117.48
n_2	Re#	D#4/Eb4	311.13	110.89
n_3	Mi	E4	329.63	104.66
n_4	Fa	F4	349.23	98.79
n_5	Fa#	F#4/Gb4	369.99	93.24
n_6	Sol	G4	392.00	88.01
n_7	Sol#	G#4/Ab4	415.30	83.07
n_8	La	A4	440.00	78.41
n_9	La#	A#4/Bb4	466.16	74.01
n_{10}	Si	B4	493.88	69.85
n_{11}	Do	C5	523.25	65.93
n_{12}	Do#	C#5/Db5	554.37	62.23

Therefore, it is possible to use mathematical operations to compute tone transposition to a different key, by adding a constant l to all elements of the series.

$$T_l(N) = T_l(n_1, n_2, n_3, n_4, n_5, n_6, n_7, n_8, n_9, n_{10}, n_{11}, n_{12})$$
$$= (n_1 + l, n_2 + l, n_3 + l, n_4 + l, n_5 + l, n_6 + l, n_7 + l,$$
$$\cdots n_8 + l, n_9 + l, n_{10} + l, n_{11} + l, n_{12} + l). \tag{3.3}$$

For example, to obtain a transposition from the key of Do (C4) to the key of Re (D4) it is necessary to add two semitones ($l = 2$) to each note of the scale, to obtain the new transposed scale shown in Table 3.4.

The inversion of a series of tones can be obtained using the same arguments, and modular arithmetic, as

$$I(N) = I(n_1, n_2, n_3, n_4, n_5, n_6, n_7, n_8, n_9, n_{10}, n_{11}, n_{12})$$
$$= (n_1, 12 - n_2, 12 - n_3, 12 - n_4, 12 - n_5, 12 - n_6, 12 - n_7,$$
$$\cdots 12 - n_8, 12 - n_9, 12 - n_{10}, 12 - n_{11}, 12 - n_{12}). \tag{3.4}$$

Retrograde inversion is a musical technique that inverts the series, that is then sounded in reverse order, that is, the first pitch becomes the last, and the last pitch turns into the first. The technique is used in twelve-tone technique (dodecaphony), a method of musical composition devised by the Austrian composer Josef Matthias Haue (1883–1959).

The retrograde inversion of the series N is obtained as

$$R(N) = R(n_1, n_2, n_3, n_4, n_5, n_6, n_7, n_8, n_9, n_{10}, n_{11}, n_{12})$$
$$= (n_{12}, n_{11}, n_{10}, n_9, n_8, n_7, n_6, n_5, n_4, n_3, n_2, n_1). \tag{3.5}$$

Mathematics can also express itself in the composition process, with the aid of self-similar and infinite automaton processes, with the use of mapping for geometric patterns, Mandelbrot's fractals, Fibonacci numbers, Cantor set, and Pascal's triangle [Johnson, 1996].

Mathematicians are fond of aesthetics, and it is commonplace to say that a mathematical deduction is beautiful, and that the solution found is elegant. Both music and mathematics use symbolic notations that simplify the composition, facilitate the creation, and help the mathematicians and musicians play in tune with their respective fellows.

Claude Debussy (1862–1918), the influential French composer, used to say that "Music is the arithmetic of sounds as optics is the geometry of light" [Arbonés and Milrud, 2012].

3.2 The First Studies of Musical Consonance

Pythagoras arrived at the idea that the number is the essence of all things, established the first rules, regarding the sections of a string, to obtain consonance between the produced sounds. He discovered the harmonic progressions in the notes of the musical scales, by finding the relation between the length of a string and the pitch of its vibrating note.

In order to study the properties of the sound, and to put it in a solid mathematical basis, Pythagoras created the monochord, that is illustrated in Figure 3.3.

In Figure 3.3, the string, tied at A, is subject to a tension caused by W, a suspended weight, and two bridges, a fixed one B and a movable bridge C, limit the vibration to a range BC, and D is a freely moving wheel [Newman, 2000a].

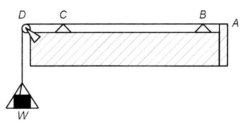

Figure 3.3   The monochord, an instrument devised by Pythagoras to study consonance. (Adapted from Hyacinth – Own work, CC BY-SA 4.0, commons.wikimedia.org/w/index.php?curid=48770590)

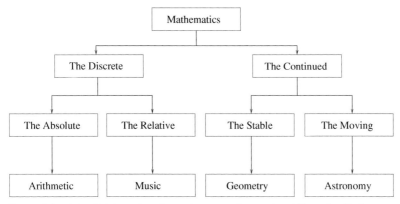

Figure 3.4 The division of mathematics, according to the philosopher Pythagoras.

Considering the same tension applied to the string, Pythagoras found that for the octave the ratio between the string lengths must be 2:1, for the fifth the ratio is 3:2, and for the fourth the ratio is 4:3 [Eves, 2011].

Those were the first registered results of mathematical physics, and they led the Pythagoreans to study musical scales, and conclude that the sound intervals were related according to the following pattern [Arbonés and Milrud, 2012]

$$f(n) = \frac{n+1}{n}, \tag{3.6}$$

in which n is an integer.

Pythagoras also coined the word mathematics and devised its fundamental branches, which can be seen in Figure 3.4 [Newman, 2000b]. For many centuries music remained a branch of discrete mathematics, and the classification is the origin of the *quadrivium*, the second part of the superior studies.

Since ancient times, mathematicians exhibit a natural affinity to music. Archytas (428 BC–347 BC), an Italian born ancient Greek philosopher mathematician, astronomer, statesman, and strategist, was a scientist of the Pythagorean school. He gave the numerical ratios for the intervals of the tetrachord on the enharmonic, the chromatic and the diatonic scales. He established that sound was due to impact, and that higher tones corresponded to faster, and lower tones to slower, motion communicated to the air [Newman, 2000b].

This is an impressive conclusion, considering that it is now widely accepted that the air molecules fill the space and suffer collisions that modify their velocity and direction, basically without loss of energy. The individual molecules velocity is of no interest, but statistically the gas pressure difference, caused by the sound, induces a compression propagation in the air with the velocity limited by the medium characteristic [Liénard, 1977].

Claudius Ptolemy (c.100–170), a Greco-Roman astronomer, mathematician, astronomer, geographer, and astrologer, analyzed some of the Pythagorean ideas and the music of the spheres. He devised a model of the solar system and also wrote the book *Concerning Harmonics*, that discussed Pythagoras's integer ratios and their connection to musical intervals.

Ptolemy, however, saw no reason to accept the number four as the highest simple integer for musical ratios. He added the integer five to create the major third (5:4), which had not been considered a consonant interval by the ancient Greeks [Johnston, 1989].

Anicius Manlius Severinus Boëthius (c.477–524), a Roman senator, consul, *magister officiorum*, and philosopher, in his book *De Institutione Musica*, divided the music into the following classes: *musica mundana*, sometimes referred to as *musica universalis*, or music of the spheres; *musica humana*, or music of the human body and soul; and *musica instrumentalis*, or vocal and instrumental music.

Musica mundana, which had existed since Pythagoras, was often taught in the *quadrivium*. Boëthius believed that arithmetic and music were interlaced, and tried to reinforce the understanding of each discipline [Grout, 1980].

3.3 Mathematicians and Music

The German astronomer Johannes Kepler (1571–1630), who studied ethics, religion, dialectics, rhetoric, physics, and astronomy, intended to recover the Pythagorean idea of the harmony of the spheres. In his book *Harmonies Mundi*, the harmony of the worlds, in 1619, he proposed that each planet produced a sound during its translation around the Sun, which depended on its angular velocity. According to his calculations, the melodies generated by Venus and the Earth varied in an interval of a semitone (half-step), or less [Arbonés and Milrud, 2012].

Gottfried Wilhelm von Leibniz (1646–1716), a German polymath and philosopher, who studied the history of mathematics and the history of philosophy, who conceived the ideas of differential and integral calculus, independently of Isaac Newton (1642–1726), used to say that "Music is the pleasure that the human mind experiments while counting, without noticing that is actually counting."

Figure 3.5 A portrait of Leonhard Euler. (Public domain. Adapted from Wikimedia Commons.)

Leonhard Euler (1707–1783), the great Swiss mathematician, physicist, astronomer, logician and engineer, tried to transform music theory into a branch of mathematics, as suggested by Pythagoras, to deduce in an orderly way, from correct principles, everything that could fit to make the mix of sounds a pleasant experience. He used to relax playing the *cravo* [du Sautoy, 2007]. A portrait of Euler is shown in Figure 3.5.

Jean le Rond D'Alembert (1717–1783), a newborn abandoned at the Saint Jean-le-Rond church, in Paris, was a mathematician, mechanician, physicist, philosopher, and music theorist. In 1752, d'Alembert wrote *Eléments de musique théorique et pratique suivant les principes de M. Rameau*, a comprehensive survey of the works of the famous French composer, writer and music theorist Jean-Philippe Rameau (1683–1764).

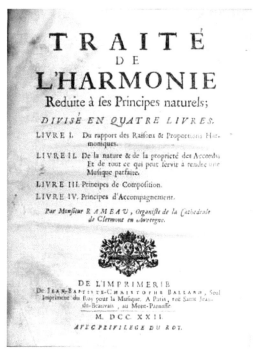

Figure 3.6   Title page of the *Traité de l'harmonie réduite à ses principes naturels*. (Public domain. Adapted from en.wikipedia.org.)

Rameau wrote a masterpiece on music theory, *Traité de l'harmonie réduite à ses principes naturels*, in 1722, as shown in Figure 3.6, and later on, in 1750, published the treatise *Démonstration du principe de l'harmonie*, in conjunction with Denis Diderot (1713–1784), a French philosopher, art critic, and writer, known as chief editor and contributor to the *Encyclopédie, ou dictionnaire raisonné des sciences, des arts et des métiers*, a general encyclopedia published in France between 1751 and 1772.

In 1746, D'Alembert reformulated the vibrating string problem, and proposed a partial differential equation, known as the one-dimensional (1D) wave equation, to solve it. At a sufficiently large distance from the sound source, the wave fronts become plane. The solutions of the wave equation then depend only on the coordinate x in the direction of propagation, and the wave equation reduces to [Maor, 2018]

$$\frac{\partial^2 u}{\partial x^2} = \frac{1}{c^2}\frac{\partial^2 u}{\partial t^2},$$

(3.7)

in which, $u = u(x, t)$ is the vertical displacement of a point on the string, positioned at a distance x from the extremities in time t. The symbols $\frac{\partial^2 u}{\partial x^2}$ and $\frac{\partial^2 u}{\partial t^2}$ are the second derivative of $u(x, t)$ with respect to x and t.

The constant c represents the wave propagation velocity on the string, and its value depends on the string tension U and on the string mass density μ,

$$c = \sqrt{\frac{U}{\mu}}, \tag{3.8}$$

in which

$$\mu = \frac{m}{l}, \tag{3.9}$$

for a string of length l and mass m.

A solution to Equation 3.7 can be proposed. For example, suppose

$$u(x, t) = \cos(\omega t - kx), \tag{3.10}$$

in which $\omega = 2\pi f$ is the angular frequency of the wave, in radians per second, $k = 2\pi/\lambda$ is the wave number, defined as the number of radians per unit distance, and λ is the wavelength.

First, consider that, for the proposed solution, the second derivative gives (Refer to Appendix A)

$$\frac{\partial^2 u}{\partial x^2} = -k^2 \cos(\omega t - kx),$$

and

$$\frac{\partial^2 u}{\partial t^2} = -\omega^2 \cos(\omega t - kx).$$

Therefore, if the results are inserted into 3.7, one obtains

$$k^2 \cos(\omega t - kx) = \frac{\omega^2}{c^2} \cos(\omega t - kx), \tag{3.11}$$

which is true if, and only if,

$$c = \frac{\omega}{k}. \tag{3.12}$$

It can be noted that the following equation is also a solution to the wave equation,

$$u(x, t) = \cos(\omega t + kx), \tag{3.13}$$

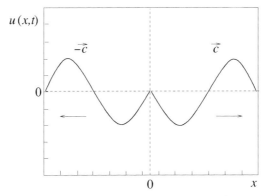

Figure 3.7   A solution to the wave equation.

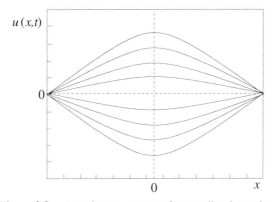

Figure 3.8   A stationary wave, such as, a vibrating string.

but, because of the linearity property of the derivative, the sum is also a solution to the wave equation. Thus, the more general solution is

$$u(x, t) = \cos\left(\omega t - kx\right) + \cos\left(\omega t + kx\right),\qquad(3.14)$$

which represents two propagating waves on the tensioned string, traveling at constant velocity c, in opposite directions, as shown in Figure 3.7.

In case there are no strings attached, the two waves will move indefinitely, in opposite directions. If the string extremities are attached to fixed points, the waves will reflect and return, forming a stationary, or standing, wave, that causes the known vibratory movement of a string guitar, for example, as shown in Figure 3.8.

Equating 3.8 and 3.12,

$$c = \sqrt{\frac{U}{\mu}} = \frac{\omega}{k}, \tag{3.15}$$

one obtains the string frequency

$$\omega = k\sqrt{\frac{U}{\mu}}. \tag{3.16}$$

Inserting Formula 3.9 into 3.16, it is possible to obtain the fundamental frequency, in radians per second,

$$\omega = k\sqrt{\frac{Ul}{m}}, \tag{3.17}$$

as a function of the wave number k, the string tension U, length l, and mass m.

The three-dimensional (3D) wave equation, to solve the wave movement in space, is given by [Loy, 2011a],

$$\frac{1}{c^2}\frac{\partial^2 u}{\partial t^2} = \frac{\partial^2 u}{\partial x^2} + \frac{\partial^2 u}{\partial y^2} + \frac{\partial^2 u}{\partial z^2}, \tag{3.18}$$

in which, $u = u(x, y, z, t)$ is the displacement of a point in space, positioned at (x, y, z), in time t.

The 3D wave equation can be simplified to

$$\frac{1}{c^2}\frac{\partial^2 u}{\partial t^2} = \nabla^2 u, \tag{3.19}$$

using the Laplacian operator

$$\nabla^2 u = \frac{\partial^2 u}{\partial x^2} + \frac{\partial^2 u}{\partial y^2} + \frac{\partial^2 u}{\partial z^2}. \tag{3.20}$$

The more general problem of how to decompose a given signal into its harmonics (partials) was solved by another great French mathematician.

3.4 The Decomposition of Music

Jean-Baptiste Joseph Fourier (1768–1830) was born to a humble family in the city of Auxerre, France, in 1768. His father was a tailor and there is little

Figure 3.9 Jean-Baptiste Joseph Fourier. (Adapted from Creative Commons. Public Domain.)

information about his mother. Both died when Fourier was still a child. He was educated, thanks to charity, to be a priest. Figure 3.9 shows a picture of Jean-Baptiste Joseph Fourier.

Fourier humble origins did not stop him from joining the army. Later, he went to school at the Superior Normal School of France, where he studied with Pierre-Simon de Laplace and Adrien-Marie Legendre. He ended up being a professor himself, discovering important mathematical theorems and contributing to the theory of differential equations.

Fourier served with Napoléon Bonaparte (1769–1821) in Egypt, where he led a research center, the Egypt Institute, and practically controlled the entire lower Nile region. He wrote a treaty on Egyptology which made him famous in the area. He became Mayor of Isère and Rhône, in France, named by Napoleon himself.

His name is associated with a series of sines and cosines, which he proposed as a solution to the differential equation related to heat conduction in solids. This solution was revolutionary for the time and was criticized by Laplace, Legendre, and Denis Poisson, demigods of mathematics.

However, Fourier is venerated by the transform which carries his name, the continuous version of the series, which is obtained when an infinite number of frequencies with infinitesimally close contributions are considered.

The Fourier transform is used to analyze signals and open up the world to another domain, beyond time, the domain of frequency. This transform is still the main worktool of every physics, mathematics, electrical, electronic and telecommunications engineering professional.

More than that, the Fourier transform is intrinsically related to the theory of probabilities, stochastic processes, source coding, error control coding, image and signal processing, harmonic structure analysis, movement of waves in the sea, electromagnetic propagation, filter projects, modulation theory, harmonics in power systems, electromagnetic pollution, just to name a few of the applications. Applications which not even the great Fourier could predict [Alencar, 2007a].

Besides the studies on heat transfer in solids, Fourier analyzed the effects in liquids and in the air. Anticipating the discussions on global warming by almost two centuries, he wrote, in 1824, printed in 1827, a long article for the Science Academy of France, titled "The temperatures of the terrestrial globe and the planetary spaces."

In this article, Fourier sought out to establish the set of phenomena and the mathematical relations between them, to explain in a general manner global warming. According to the article, the warmth of the globe derives from three distinct sources.

Earth is heated by the solar rays in a non-uniform manner, causing climate diversity. The planet is submitted to the common temperature of the planetary spaces, being exposed to the irradiation of the uncountable stars which exist in all the parts of the solar system. Lastly, Earth preserved in its interior a part of the primitive heat, which it holds since the time the planets were formed.

In particular, the rays that the Sun sends incessantly to the terrestrial globe produce two very distinctive effects. One is periodical and encompasses basically the exterior envelopment of the Earth. This effect consists of daily or annual climate variations. The other is constant and can be observed in deep places, for instance, well below the surface.

The presence of the atmosphere and water makes for a more uniform heat distribution. According to Fourier, the Sun rays that reach the Earth in the form of light have the property of penetrating solid or liquid substances.

When reaching these earthly bodies, these rays become an obscure radiant heat, called infrared radiation, which was not known by this name at the time.

The distinction between luminous heat and obscure heat would explain the elevation in temperature caused by the transparent bodies, since the rays of light would easily cross the atmosphere, while the obscure rays would have a hard time accomplishing the opposite path. This effect would be responsible for the warming of the terrestrial surface. Fourier was, once again, right [Alencar, 2007b].

According to Fourier, the movement of the air and the waters, the regime of sea, and the elevation and the shape of the soil, the effects of human industry and all the accidental alterations of the terrestrial surface modify the temperatures in each climate. The presence of clouds, which intercept the rays, tempers this climate.

The largest concentration of gases in proximity to the earthly surface would cause the heat to be contained in that region. In particular, it is important that the clouds have a white color, because then they reflect the solar light and balance the rain regime and the temperatures on the surface. If the clouds were another color, or were black, life would not exist on Earth.

The mobility of the waters and the air tends to moderate the effects of heat and cold and makes the distribution more uniform. But it would be impossible, according to Fourier, for the action of the atmosphere to supplant the cause which maintains the common temperature in the planetary spaces. The other planets would have similar climatic effects, due to the gases that composed their atmospheres.

The Swedish chemist Svante Arrhenius (1859–1927), winner of the Nobel chemistry prize, in 1903, built upon Fourier's ideas and formulated the hypothesis that the human being, because of industrialization, could influence the climate by the increasing emission of carbon dioxide, one of the gases which cause the famous greenhouse effect.

Arrhenius investigated what would happen to the global climate if the concentration of carbonic gas continued increasing. However, only in 1957, the researchers Roger Revelle and Hans Suess presented, in an article, convincing evidence on the behavior of carbon dioxide produced by humanity. They demonstrated that the oceans could not absorb, as was thought at the time, all the carbonic gas that was produced.

It is interesting to note that an explanation of the slow heat absorption by the oceans had already been given by Fourier in 1824, related to his study on the transmission of heat in liquids.

It is estimated that the quantity of carbonic gas stored by the oceans reaches 36 thousand Gt (gigatons), which is 50 times the total carbon dioxide present in the atmosphere. According to Revelle and Suess, the superficial layer of the oceans absorb this gas slowly, which happens by diffusion and subsequent mixture of water layers. This last process takes years or centuries.

If the oceans can not sequester the carbonic gas from the air easily, and the forests have a balanced process of emission and absorption, this results in a gradual accumulation of carbon dioxide in the atmosphere. This accumulation would imply the increase of average temperature on the surface of the Earth, known as global warming. This is happening in a short geological time, from tens to hundreds of years [Alencar, 2007c].

The greenhouse effect by itself is not harmful. This warming was the phenomenon which, by making it possible to reach a more adequate climate to the stabilization of human masses, permitted the burgeoning of what is known today as civilization. All the great creations of humanity, such as writing, tools, great constructions, agriculture, cities, mathematics, physics, biology, philosophy, engineering, music, were forged in the last 10 thousand years during the last period of global warming.

Exactly the time in which the Earth has kept heated to a sufficient temperature to propitiate the emergence and comfortable maintenance of life in a good share of its surface. And this time period is small, when compared to the age of the Earth, estimated at four billion years.

In a few hundred or thousand years, planet Earth will inevitably be in a new Ice Age, because these ages are periodic like the Fourier series, or more appropriately, they are cyclestationary, taking into account their random nature.

The global cooling cycles occur in intervals of approximately 11,000 years, and the next cycle should occur very soon, at least in terms of geological eras.

In the future, when it finally happen, the global temperature will fall incredibly and a good portion of the planet will have a climate identical to what can be seen in the Arctic and Antarctic poles currently [Alencar, 2007d].

4

Signal Analysis

---

*"One ought, every day at least, to hear a little song, read a good poem, see
a fine picture, and, if it were possible, to speak a few reasonable words."*
Johann Wolfgang von Goethe

4.1 Basic Fourier Analysis

This chapter provides the necessary mathematical basis for the reader to deal
with Fourier theory, and apply the concepts to understand the spectrum analysis in music. [Alencar, 1999]. The main concepts associated with Fourier
series and Fourier transform are introduced. The theory and the properties of
Fourier transform are presented, along with the main properties of signal and
system analysis [Papoulis, 1983].

Fourier theory establishes fundamental conditions for the representation
of an arbitrary function in a finite interval as a sum of sinuoids. In fact, this is
just an instance of the more general Fourier representation of signals in which
a periodic signal $f(t)$, under fairly general conditions, can be represented by
a complete set of orthogonal functions [Baskakov, 1986].

By a complete set $\mathcal{F}$ of orthogonal functions, it is understood that, except
for those orthogonal functions already in $\mathcal{F}$, there are no other orthogonal
functions that belong to $\mathcal{F}$ to be considered.

The periodic signal $f(t)$ must satisfy the Dirichlet conditions, that is, $f(t)$
is a bounded function which in any one period has at most a finite number
of local maxima and minima and a finite number of points of discontinuity
[Wylie, 1966].

The expansion of a periodic signal $f(t)$ as a sum of mutually orthogonal
functions requires a review of the concepts of periodicity and orthogonality.

A given function $f(t)$ is periodic, of period T, as illustrated in Figure 4.1,
if and only if, T is the smallest positive number for which $f(t + T) = f(t)$.

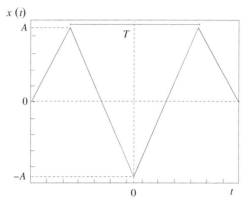

Figure 4.1   Example of a periodic signal.

In other words, $f(t)$ is periodic if its domain contains $t + T$ whenever it contains t, and $f(t + T) = f(t)$.

It follows from the definition of a periodic function that if T represents the period of $f(t)$, then $f(t) = f(t + nT)$, for $n = 1, 2, \ldots$, that is, $f(t)$ repeats its values when integer multiples of T are added to its argument [Wozencraft and Jacobs, 1965].

If $f(t)$ and $g(t)$ are two periodic functions with the same period T, then their sum $f(t) + g(t)$ will also be a periodic function with period T. This result can be proven if one makes

$$h(t) = f(t) + g(t), \tag{4.1}$$

and further notices that

$$h(t + T) = f(t + T) + g(t + T) = f(t) + g(t) = h(t). \tag{4.2}$$

Orthogonality provides the tool to introduce the concept of a basis, that is, a minimum set of functions that can be used to generate other functions. However, orthogonality by itself does not guarantee that a complete vector space is generated.

Two real functions $u(t)$ and $v(t)$, defined in the interval $\alpha \le t \le \beta$, are orthogonal if their inner product is null in the interval, that is, if

$$<u(t), v(t)> = \int_{\alpha}^{\beta} u(t)v(t)dt = 0. \tag{4.3}$$

4.1.1 The Trigonometric Fourier Series

The trigonometric Fourier series representation of the signal $f(t)$, that is, its decomposition into sinusoidal functions, can be written as

$$f(t) = a_0 + \sum_{n=1}^{\infty} [a_n \cos(n\omega_0 t) + b_n \sin(n\omega_0 t)], \tag{4.4}$$

in which the term a_0, the average value of the function $f(t)$, indicates whether or not the signal contains a constant value and the terms a_n and b_n are called Fourier series coefficients, in which n is a positive integer.

The equality sign holds in (4.4) for all values of t only when $f(t)$ is periodic. Figure 4.2 depicts examples of sine and cosine functions.

Fourier series representation is a useful tool for any type of signal, as long as that signal representation is required only in the $[0, T]$ interval. Outside that interval, Fourier series representation is always periodic, even if the signal $f(t)$ is not periodic [Knopp, 1990].

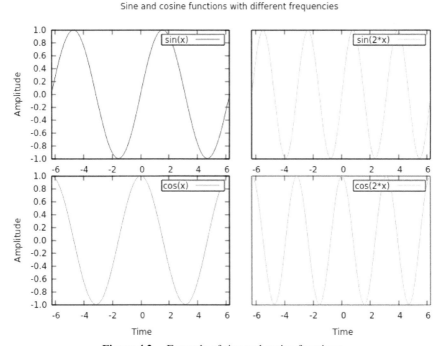

Figure 4.2 Example of sine and cosine functions.

The trigonometric functions, for instance, sine and cosine, are examples of orthogonal functions because they satisfy the following equations, called orthogonality relations, for integer values of n and m:

$$\int_0^T \cos(n\omega_o t)\sin(m\omega_o t)dt = 0, \quad \text{for all integers } n, m, \tag{4.5}$$

$$\int_0^T \cos(n\omega_o t)\cos(m\omega_o t)dt = \begin{cases} 0 & \text{if } n \neq m \\ \frac{T}{2} & \text{if } n = m \end{cases} \tag{4.6}$$

$$\int_0^T \sin(n\omega_o t)\sin(m\omega_o t)dt = \begin{cases} 0 & \text{if } n \neq m \\ \frac{T}{2} & \text{if } n = m \end{cases} \tag{4.7}$$

in which $\omega_0 = 2\pi/T$.

As a consequence of the orthogonality conditions, explicit expressions for the coefficients a_n and b_n of Fourier trigonometric series can be computed. By integrating both sides in expression (4.4) in the interval $[0, T]$, it follows that [Oberhettinger, 1990]

$$\int_0^T f(t)dt = \int_0^T a_o dt + \sum_{n=1}^\infty \int_0^T a_n \cos(n\omega_o t)dt + \sum_{n=1}^\infty \int_0^T b_n \sin(n\omega_o t)dt$$

and since

$$\int_0^T a_n \cos(n\omega_o t)dt = \int_0^T b_n \sin(n\omega_o t)dt = 0,$$

it follows that

$$a_o = \frac{1}{T}\int_0^T f(t)dt. \tag{4.8}$$

Multiplication of both sides in expression (4.4) by $\cos(m\omega_o t)$, and integrating in the interval $[0, T]$, leads to

$$\int_0^T f(t)\cos(m\omega_o t)dt = \int_0^T a_o \cos(m\omega_o t)dt \tag{4.9}$$

$$+ \sum_{n=1}^\infty \int_o^T a_n \cos(n\omega_o t)\cos(m\omega_o t)dt$$

$$+ \sum_{n=1}^\infty \int_o^T b_n \cos(m\omega_o t)\sin(n\omega_o t)dt,$$

which, after simplification, gives

$$a_n = \frac{2}{T} \int_0^T f(t) \cos(n\omega_o t) dt, \quad \text{for } n = 1, 2, 3, \ldots \tag{4.10}$$

In a similar way, b_n is found by multiplying both sides in Expression 4.4 by $\sin(n\omega_o t)$ and integrating in the interval $[0, T]$, that is,

$$b_n = \frac{2}{T} \int_0^T f(t) \sin(n\omega_o t) dt, \tag{4.11}$$

for $n = 1, 2, 3, \ldots$.

A considerable simplification, when computing coefficients of a trigonometric Fourier series, is obtained with properties of even and odd functions.

If $f(t)$ is an even function, then $b_n = 0$, and

$$a_n = \frac{2}{T} \int_0^T f(t) \cos(n\omega_o t) dt, \quad \text{for } n = 1, 2, 3, \ldots. \tag{4.12}$$

If $f(t)$ is an odd function, then $a_n = 0$ and

$$b_n = \frac{2}{T} \int_0^T f(t) \sin(n\omega_o t) dt, \quad \text{for } n = 1, 2, 3, \ldots. \tag{4.13}$$

Example: Compute the coefficients of the trigonometric Fourier series for the waveform

$$f(t) = A[u(t + \tau) - u(t - \tau)], \tag{4.14}$$

which repeats itself with period $T = 1/f$, in which $u(t)$ denotes the unit step function.

Solution: Since the given signal is symmetric with respect to the ordinate axis, it follows that $f(t) = f(-t)$ and the function is even. Therefore, $b_n = 0$, and all that is left to compute is a_o, and a_n for $n = 1, 2, \ldots$.

The expression to calculate the average value a_o is

$$a_o = \frac{1}{T} \int_{-\frac{T}{2}}^{\frac{T}{2}} f(t) dt = \frac{1}{T} \int_{-\tau}^{\tau} A dt = \frac{2A\tau}{T}.$$

In the previous equation, the maximum value of τ is $T/2$. The coefficients a_n for $n = 1, 2, \ldots$ are computed as

$$a_n = \frac{2}{T} \int_0^T f(t) \cos(n\omega_o t) dt = \frac{2}{T} \int_{-\tau}^{\tau} A \cos(n\omega_o t) dt,$$

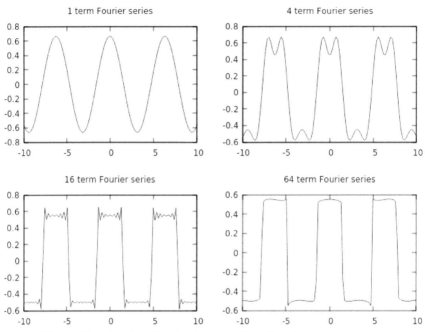

Figure 4.3   A series of cosine functions that add to form a square wave.

$$a_n = \frac{4A}{T} \int_0^\tau \cos(n\omega_o t)dt$$

$$= \frac{4A}{Tn\omega_o} \sin(n\omega_o t) \Big|_0^\tau = (4A\tau/T)\frac{\sin(n\omega_o\tau)}{n\omega_0\tau} \ .$$

Therefore, the signal $f(t)$ is represented by the following trigonometric Fourier series

$$f(t) = \frac{2A\tau}{T} + \left(\frac{4A\tau}{T}\right) \sum_{n=1}^{\infty} \frac{\sin(n\omega_o\tau)}{n\omega_0\tau} \cos(n\omega_o t).$$

Figure 4.3 shows a pictorial demonstration of a series of cosine functions, with one, four, 16 and 64 terms, that add to form a square wave. Notice that the approximation becomes more accurate as the number of terms of Fourier series increases.

Nonetheless, the representation of signals by orthogonal functions usually presents an error, which diminishes as the number of component terms in the corresponding series is increased. This error produces Gibbs phenomenon, an oscillation that occurs at the transition points [Schwartz and Shaw, 1975].

The phenomenon was discovered by Henry Wilbraham (1825–1883), an English mathematician, and rediscovered by Josiah Willard Gibbs (1839–1903), an American mathematician and physicist. It is the characteristic, and oscillatory, way in which Fourier series of a piecewise continuously differentiable periodic function behaves at a steplike discontinuity.

It is not really necessary that the signal be a periodic function to have a Fourier series decomposition. Any signal, under a few mathematical constraints, can be written as a series of sines and cosines, as long as, it is only observed in a window equivalent to its domain. Therefore, every function can be uniquely represented in an interval by a Fourier series.

This fact is known as Carleson's theorem, proved by Lennart Axel Edvard Carleson (1928–), a Swedish mathematician in the field of harmonic analysis, who established the precise conditions for pointwise almost everywhere convergence of Fourier series. The following example illustrates the point.

Example: When a string is plucked, for instance, when someone is playing an acoustic or electric guitar, the initial shape is similar to the one presented in Figure 4.4. After the string is released, it begins to vibrate in different modes, to generate the corresponding harmonics that compose the guitar *timbre*. Compute the Fourier series corresponding to the triangular signal.

Solution: The triangular wave can be represented in Fourier series, assuming that the wave repeats itself indefinitely, as shown in Figure 4.5.

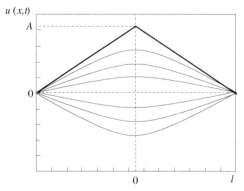

Figure 4.4   Example of a triangular wave.

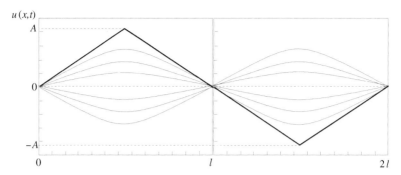

Figure 4.5   Extension of a triangular wave.

Considering that the initial wave is triangular of length $2l$, and that it is an odd function, then the coefficients of the cosine terms are null, $a_0 = 0$ and $a_n = 0$. Therefore, it is only necessary to compute the sine coefficients.

$$
b_n = \frac{2A}{l} \int_0^{l/2} \frac{x}{l/2} \sin\left(\frac{n\pi x}{l}\right) dx
$$
$$
+ \frac{2A}{l} \int_{l/2}^{l} \left[1 - \frac{2}{l}\left(x - \frac{l}{2}\right)\right] \sin\left(\frac{n\pi x}{l}\right) dx, \qquad (4.15)
$$

which gives

$$
b_n = \begin{cases} \frac{8A}{\pi^2 n^2}(-1)^{(n-1)/2}, & \text{for } n \text{ even,} \\ 0, & \text{otherwise.} \end{cases} \qquad (4.16)
$$

Some time after the string is released, a stationary sine wave is formed, of the type $u(x, t) = \sin(\omega t - kx)$. If a node appears at $x = l$, then

$$
u(l, t) = \sin(\omega t - kl) = 0,
$$

therefore,

$$
\omega t - kl = 0,
$$

or

$$
\frac{\omega}{k} = \frac{l}{t} = c,
$$

in which c is the sound velocity, $k = 2\pi/\lambda$ is the wave number, and $\omega = 2\pi f$, is the angular frequency.

Then, one obtains

$$
c = \frac{2\pi f}{2\pi/\lambda} = \lambda f,
$$

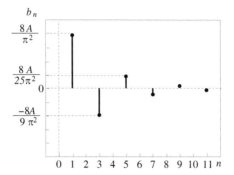

Figure 4.6 Spectral coefficients of a triangular wave.

which finally gives a well-known relation between the wavelength, the sound velocity and the frequency,

$$f = \frac{c}{\lambda}.$$

Finally, the triangular function can be expressed as a Fourier series, for $t = 0$, as follows

$$u(x,0) = \frac{8A}{\pi^2} \sum_{n=1,3,5,\dots}^{\infty} \frac{(-1)^{(n-1)/2}}{n^2} \sin\left(\frac{n\pi x}{l}\right). \tag{4.17}$$

The Fourier coefficients form an alternating and decreasing series, for odd harmonics (partials), and are depicted in Figure 4.6.

4.1.2 The Compact Fourier Series

It is possible to represent Fourier series in a form known as the compact Fourier series, as follows

$$f(t) = C_0 + \sum_{n=1}^{\infty} C_n \cos(n\omega_o t + \theta_n). \tag{4.18}$$

By expanding the expression

$$C_n \cos(n\omega_o t + \theta) = C_n \cos(n\omega_o t) \cos\theta_n - C_n \sin(n\omega_o t) \sin\theta_n, \tag{4.19}$$

and comparing this result with 4.4, it follows that

$$a_o = C_o, \tag{4.20}$$

$$a_n = C_n \cos \theta_n, \tag{4.21}$$

and

$$b_n = -C_n \sin \theta_n. \tag{4.22}$$

It is now possible to compute C_n as a function of a_n and b_n. For that purpose, it is sufficient to square a_n and b_n and add the result, that is,

$$a_n^2 + b_n^2 = C_n^2 \cos^2 \theta_n + C_n^2 \sin^2 \theta_n = C_n^2. \tag{4.23}$$

From Equation 4.23, the modulus of C_n can be written as

$$C_n = \sqrt{a_n^2 + b_n^2}. \tag{4.24}$$

In order to determine θ_n, it suffices to divide b_n by a_n, that is,

$$\frac{b_n}{a_n} = -\frac{\sin \theta_n}{\cos \theta_n} = -\tan \theta_n, \tag{4.25}$$

which, when solved for θ_n, gives

$$\theta_n = -\arctan\left(\frac{b_n}{a_n}\right). \tag{4.26}$$

4.1.3 The Exponential Fourier Series

Since the set of exponential functions can be written, using Euler's identity, as

$$e^{jn\omega_o t} = \cos(n\omega_o) + j\sin(n\omega_o), \quad n = 0, \pm 1, \pm 2, \ldots, \tag{4.27}$$

it is recognized as a complete set of orthogonal functions in an interval of magnitude T, in which $T = 2\pi/\omega_o$.

Therefore, it is possible to represent a function $f(t)$ by a linear combination of exponential functions in an interval T,

$$f(t) = \sum_{-\infty}^{\infty} F_n e^{jn\omega_o t}, \tag{4.28}$$

in which

$$F_n = \frac{1}{T} \int_{-\frac{T}{2}}^{\frac{T}{2}} f(t) e^{-jn\omega_o t} dt. \tag{4.29}$$

Equation 4.28 represents the exponential Fourier series expansion of $f(t)$ and Equation 4.29 is the expression to compute the associated series coefficients.

The exponential Fourier series is also known as the complex Fourier series. It can be shown that Equation 4.28 is just another way of expressing the Fourier series as given in 4.4.

Replacing $\cos(n\omega_o t) + j\sin(n\omega_o t)$ for $e^{n\omega_o t}$, in (4.28), it follows that

$$f(t) = F_o + \sum_{n=-\infty}^{-1} F_n[\cos(n\omega_o t) + j\sin(n\omega_o t)]$$

$$+ \sum_{n=1}^{\infty} F_n[\cos(n\omega_o t) + j\sin(n\omega_o t)],$$

or

$$f(t) = F_o + \sum_{n=1}^{\infty} F_n[\cos(n\omega_o t) + j\sin(n\omega_o t)] + F_{-n}[\cos(n\omega_o t) - j\sin(n\omega_o t)].$$

After grouping the coefficients of the sine and cosine terms, one obtains

$$f(t) = F_o + \sum_{n=1}^{\infty} (F_n + F_{-n})\cos(n\omega_o t) + j(F_n - F_{-n})\sin(n\omega_o t). \quad (4.30)$$

Comparing this expression with Expression 4.4, it follows that

$$a_o = F_o, \quad a_n = (F_n + F_{-n}) \quad \text{and} \quad b_n = j(F_n - F_{-n}), \quad (4.31)$$

and that

$$F_o = a_o, \quad (4.32)$$

$$F_n = \frac{a_n - jb_n}{2}, \quad (4.33)$$

and

$$F_{-n} = \frac{a_n + jb_n}{2}. \quad (4.34)$$

In case the function $f(t)$ is even, that is, if $b_n = 0$, then

$$a_o = F_o, \quad F_n = \frac{a_n}{2}, \quad \text{and} \quad F_{-n} = \frac{a_n}{2}. \quad (4.35)$$

Example: Compute the exponential Fourier series for the train of impulses given by,

$$\delta_T(t) = \delta[\sin(2\pi f t)], \quad f = 1/T.$$

Solution: The complex coefficients are given by

$$F_n = \frac{1}{T} \int_{\frac{-T}{2}}^{\frac{T}{2}} \delta_T(t) e^{-jn\omega_o t} dt = \frac{1}{T}, \tag{4.36}$$

using the property of impulse filtering,

$$\int_{-\infty}^{\infty} \delta(t - t_o) f(t) dt = f(t_o). \tag{4.37}$$

Observe that $f(t)$ can be written as

$$f(t) = \frac{1}{T} \sum_{n=-\infty}^{\infty} e^{-jn\omega_o t}. \tag{4.38}$$

The impulse train is an idealization, as most functions are, of a real signal. In practice, in order to obtain an impulse train, it is sufficient to pass a binary digital signal through a differentiator circuit and then pass the resulting waveform through a half-wave rectifier.

Fourier series expansion of a periodic signal is equivalent to its decomposition in frequency components. In general, a periodic function with period T has frequency components $0, \pm\omega_o, \pm2\omega_o, \pm3\omega_o, \dots, \pm n\omega_o$, in which $\omega_o = 2\pi/T$ is the fundamental frequency and the multiples of ω_0 are called harmonics. Notice that the spectrum exists only for discrete values of ω and that the spectral components are spaced by at least ω_o.

4.2 Convergence to Fourier Transform

It has been shown that an arbitrary function can be represented in terms of an exponential (or trigonometric) Fourier series in a finite interval. If such a function is periodic, this representation can be extended for the entire real interval $(-\infty, \infty)$.

However, it is interesting to observe the spectral behavior of a function in general, periodic or not, in the entire interval $(-\infty, \infty)$. To do that, it is necessary to truncate the function $f(t)$ in the interval $[-T/2, T/2]$, to obtain $f_T(t)$.

It is possible, then, to represent this function as a sum of exponentials in the entire interval $(-\infty, \infty)$ if T goes to infinity, as follows.

$$\lim_{T \to \infty} f_T(t) = f(t).$$

The signal $f_T(t)$ can be represented by the exponential Fourier series, as

$$f_T(t) = \sum_{n=-\infty}^{\infty} F_n e^{jn\omega_o t}, \tag{4.39}$$

in which $\omega_o = 2\pi/T$ and

$$F_n = \frac{1}{T} \int_{-\frac{T}{2}}^{\frac{T}{2}} f_T(t) e^{-jn\omega_o t} dt. \tag{4.40}$$

The coefficients F_n represent the spectral amplitude associated to each component of frequency $n\omega_o$.

As T increases, the amplitudes diminish but the spectrum shape is not altered. The increase in T forces ω_o to diminish and the spectrum to become denser.

In the limit, as $T \to \infty$, ω_o becomes infinitesimally small, being represented by $d\omega$. On the other hand, there are now infinitely many components and the spectrum is no longer a discrete one, becoming a continuous spectrum in the limit.

For convenience, write $TF_n = F(\omega)$, that is, the product TF_n becomes a function of the variable ω, since $n\omega_o \to \omega$. Replacing $\frac{F(\omega)}{T}$ for F_n in 4.39, one obtains

$$f_T(t) = \frac{1}{T} \sum_{n=-\infty}^{\infty} F(\omega) e^{j\omega t}. \tag{4.41}$$

Replacing $\omega_0/2\pi$ for $1/T$,

$$f_T(t) = \frac{1}{2\pi} \sum_{n=-\infty}^{\infty} F(\omega) e^{j\omega t} \omega_0. \tag{4.42}$$

In the limit, as T approaches infinity, one has

$$f(t) = \frac{1}{2\pi} \int_{-\infty}^{\infty} F(\omega) e^{j\omega t} d\omega \tag{4.43}$$

which is known as the inverse Fourier transform.

Similarly, from 4.40, as T approaches infinity, one obtains

$$F(\omega) = \int_{-\infty}^{\infty} f(t) e^{-j\omega t} dt \tag{4.44}$$

which is known as the direct Fourier transform, sometimes written as $F(\omega) = \mathcal{F}[f(t)]$. A Fourier transform pair is often denoted as

$$f(t) \longleftrightarrow F(\omega).$$

While correcting his work on the analysis of heat transfer, Fourier tried to understand the graphics that represented the physical phenomena, for example, the temperature evolution with time, or the meaning of a sound wave. He knew that the sound could be represented by a plot, in which the horizontal axis was associated with time, and the vertical axis with the sound intensity.

Fourier began to explore the way complex sounds could be produced using pure sine waves. He imagined that several tuning forks, played at the same time, could emulate, for instance, the sound produced by an orchestra [du Sautoy, 2007].

5

Fourier Transform Applications

"There is geometry in the humming of the strings, there is music in the spacing of the spheres."

Pythagoras

5.1 Fourier Transform

This chapter deals with applications of Fourier transform to the analysis of sound and music parameters. The properties of Fourier transform are presented, along with the main properties of signal and system analysis [Papoulis, 1983].

Signal analysis is an important tool to understand music, and to know the sound captured by a set of microphones is sampled, quantized, and digitized, to be stored in a compact disc (CD), or to be sent over the Internet. The direct Fourier transform, derived in Chapter 4,

$$F(\omega) = \int_{-\infty}^{\infty} f(t)e^{-j\omega t}dt \tag{5.1}$$

and the inverse Fourier transform,

$$f(t) = \frac{1}{2\pi} \int_{-\infty}^{\infty} F(\omega)e^{j\omega t}d\omega \tag{5.2}$$

form the basis or modern signal analysis [Haykin, 1988]. Some important Fourier transforms are presented in the following [Alencar and da Rocha Jr., 2005].

5.1.1 Fourier Transform of the Impulse Function

The impulse function $\delta(t)$ is a mathematical model of a very strong hit on a percussion instrument, such as a drum, as shown in Figure 5.1.

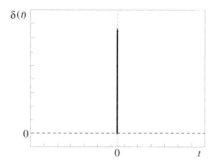

Figure 5.1   Example of an impulse function.

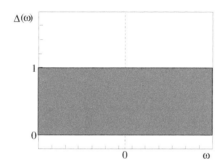

Figure 5.2   Spectrum of the impulse function.

In Formula 4.44, by making the substitution $f(t) = \delta(t)$, it follows that

$$\Delta(\omega) = \int_{-\infty}^{\infty} \delta(t)e^{-j\omega t}dt. \tag{5.3}$$

Using the impulse filtering property, one obtains $\Delta(\omega) = 1$. Because the Fourier transform is constant, over the entire spectrum, one concludes that the impulse function contains a continuum of equal amplitude spectral components. The spectrum of the impulse function is shown in Figure 5.2.

An impulse can model a sudden electrostatic discharge, such as the lightning that occurs during a thunderstorm. The effect of the lightning can be felt, for example, in several radiofrequency ranges, including the amplitude modulation (AM) and frequency modulation (FM) radio bands, and the television (TV) band.

Alternatively, by making $F(\omega) = 1$ in 4.43 and simplifying, the impulse function can be written as

$$\delta(t) = \frac{1}{\pi} \int_{0}^{\infty} \cos \omega t \, d\omega.$$

5.1.2 Transform of the Unit Step Function

The unit step function is the integral of the impulse function

$$u(t) = \int_{\infty}^{t} \delta(\tau) d\tau, \tag{5.4}$$

and models a signal that has been turned on at time zero. It is useful to limit the operation range of certain signals, and is illustrated in Figure 5.3.

Example: A periodic square wave $s(t)$, shown in Figure 5.4, can be represented, using the unit step function, as follows

$$s(t) = u[\cos(\omega t)],$$

in which $\omega = 2\pi f$ is an angular frequency, in radians per second. The first zero of the cosine function is obtained by setting $\omega t = \pi/2$. Therefore, the period of the square wave is $T = 2\pi/\omega$.

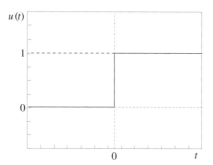

Figure 5.3   The unit step function.

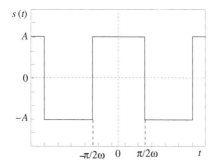

Figure 5.4   A square wave.

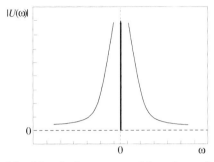

Figure 5.5   Magnitude spectrum of the unit step function.

The Fourier transform of the unit step function is

$$U(\omega) = \pi\delta(\omega) + \frac{1}{j\omega}. \tag{5.5}$$

The spectrum of the unit step function is depicted in Figure 5.5. It is interesting to note that the spectrum is composed of an impulse at the origin, caused by the constant part of the function, and by decreasing frequency amplitudes that range from zero to infinity, caused by the jump at the origin.

The unit step function can be represented, in integral form, as

$$u(t) = \frac{1}{\pi} \int_0^\infty \frac{\sin(\omega t)}{\omega} d\omega.$$

Is is possible to show that $u(t) * u(t) = r(t)$, in which $u(t)$ is the unit step function, $r(t)$ denotes the ramp function, and the symbol $*$ denotes the convolution operation.

5.1.3 Transform of the Constant Function

A word of caution is necessary in this case. If $f(t) = A$ is a constant function, as shown in Figure 5.6, then its Fourier transform in principle would not exist since this function does not satisfy the absolute integrability criterion.

In general, $F(\omega)$, Fourier transform of $f(t)$, is expected to be finite, that is,

$$|F(\omega)| \leq \int_{-\infty}^{\infty} |f(t)||e^{-j\omega t}|dt < \infty, \tag{5.6}$$

since $|e^{-j\omega t}| = 1$, then

$$\int_{-\infty}^{\infty} |f(t)|dt < \infty. \tag{5.7}$$

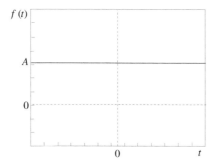

Figure 5.6 A constant function.

However, that is just a sufficiency condition and not a necessary one for the existence of the Fourier transform, since there exist functions that although do not satisfy the condition of absolute integrability, in the limit have a Fourier transform [Carlson, 1975].

This observation is important, since this approach is often used to compute Fourier transforms of several functions. The constant function can be approximated by a gate function with amplitude A and width τ, if τ is increased to approach infinity,

$$\mathcal{F}[A] = \lim_{\tau \to \infty} A\tau \mathrm{Sa}\left(\frac{\omega\tau}{2}\right) \tag{5.8}$$

$$= 2\pi A \lim_{\tau \to \infty} \frac{\tau}{2\pi} \mathrm{Sa}\left(\frac{\omega\tau}{2}\right),$$

then,

$$\mathcal{F}[A] = 2\pi A\delta(\omega). \tag{5.9}$$

This result, depicted in Figure 5.7, is interesting and also intuitive, since a constant function in time represents a direct current (DC) level and, as was to be expected, contains no spectral component except for the one at $\omega = 0$.

5.1.4 The Exponential Signal

An exponential function is denoted as

$$f(t) = e^{-at}u(t), \tag{5.10}$$

and is shown in Figure 5.8. It can model the sound envelope of an instrument whose attack period is very short, and the remaining time is spent on the release period.

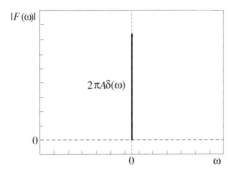

Figure 5.7   The spectrum of a constant function.

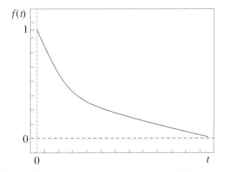

Figure 5.8   Example of an exponential function.

The exponential function is also useful to model the latency in computer networks, the probability distribution of letters in an alphabet, and the arrival of phone calls in a telephone system.

It follows from Formula 4.44 that Fourier transform of the exponential function is given by

$$F(\omega) = \int_{-\infty}^{\infty} e^{-at}u(t)e^{-j\omega t}dt = \int_{0}^{\infty} e^{-at}e^{-j\omega t}dt \qquad (5.11)$$

$$= \int_{0}^{\infty} e^{-(a+j\omega)t}dt = \left[-\frac{e^{-(a+j\omega)t}}{a+j\omega} \right]_{0}^{\infty},$$

which gives,

$$F(\omega) = \frac{1}{a+j\omega}. \qquad (5.12)$$

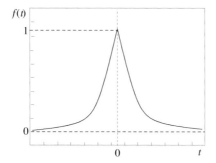

Figure 5.9 Example of a bilateral exponential function.

The spectrum magnitude is given by

$$|F(\omega)| = \frac{1}{\sqrt{a^2 + \omega^2}}. \tag{5.13}$$

5.1.5 Bilateral Exponential Signal

A bilateral exponential function is denoted as

$$f(t) = e^{-a|t|}, \tag{5.14}$$

and shown in Figure 5.9. This function models, for instance, the probability distribution of the voice signal, and can be used as a more realistic approximation for the impulse function.

It follows from Formula 4.44 that the Fourier transform of the bilateral exponential function is given by

$$
\begin{aligned}
F(\omega) &= \int_{-\infty}^{\infty} e^{-a|t|} e^{-j\omega t} dt \\
&= \int_{-\infty}^{0} e^{at} e^{-j\omega t} dt + \int_{0}^{\infty} e^{-at} e^{-j\omega t} dt \\
&= \frac{1}{a - j\omega} + \frac{1}{a + j\omega}, \tag{5.15}
\end{aligned}
$$

$$F(\omega) = \frac{2a}{a^2 + \omega^2}. \tag{5.16}$$

Because Fourier transform is a real function, it can be plotted directly, without the need to compute the magnitude, as shown in Figure 5.10.

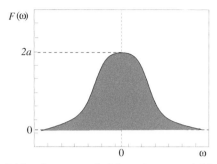

Figure 5.10   Spectrum of a bilateral exponential function.

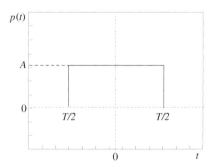

Figure 5.11   Example of a gate function.

5.1.6 Transform of the Gate Function

The gate function, useful as a mathematical model for the sound envelope of a simple electronic synthesizer, is shown in Figure 5.11. The gate function can be represented in terms of the unit step function $u(t)$, as follows

$$p(t) = A[u(t + T/2) - u(t - T/2)]. \tag{5.17}$$

Fourier transform of the gate function, can be calculated as follows,

$$\begin{aligned}
P(\omega) &= \int_{-\frac{T}{2}}^{\frac{T}{2}} Ae^{-j\omega t}dt \\
&= \frac{A}{j\omega}(e^{j\omega\frac{T}{2}} - e^{-j\omega\frac{T}{2}}) \\
&= \frac{2jA}{j\omega}\sin(\omega T/2), \tag{5.18}
\end{aligned}$$

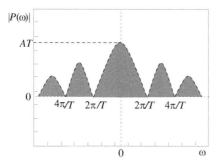

Figure 5.12 Magnitude plot of the Fourier transform of the gate function.

which can be rearranged as

$$P(\omega) = AT \left(\frac{\sin(\omega T/2)}{\omega T/2} \right),$$

and finally

$$P(\omega) = AT \text{Sa} \left(\frac{\omega T}{2} \right), \tag{5.19}$$

in which

$$\text{Sa}(x) = \frac{\sin x}{x} \tag{5.20}$$

is the sampling function. This function converges to one, as x goes to zero, and converges to zero, as x goes to infinity.

The sampling function, the magnitude of which is illustrated in Figure 5.12, is of great relevance in communication theory, signal processing, optics and sound theory.

The sampling function obeys the following important relationship

$$\frac{k}{\pi} \int_{-\infty}^{\infty} \text{Sa}(kt) dt = 1. \tag{5.21}$$

The area under this curve is equal to 1. As k increases, the amplitude of the sampling function increases, the spacing between zero crossings diminishes, and most of the signal energy concentrates near the origin.

For $k \to \infty$, the sampling function converges to the impulse function,

$$\delta(t) = \lim_{k \to \infty} \frac{k}{\pi} \text{Sa}(kt). \tag{5.22}$$

In this manner, in the limit, it is true that $\int_{-\infty}^{\infty} \delta(t)dt = 1$. Since the function concentrates its non-zero values near the origin, it follows that $\delta(t) = 0$ for $t \neq 0$.

Equation 5.22, which is a limiting property of the sampling function, can be used as a definition of the impulse. In fact, several functions that have unit area, such as the Gaussian function, can also be used to define the impulse.

5.1.7 Fourier Transform of the Sine and Cosine Functions

Since both the sine and the cosine functions are periodic functions, they do not satisfy the condition of absolute integrability. However, their respective Fourier transforms exist in the limit.

Assuming the function to exist only in the interval $(\frac{-\tau}{2}, \frac{\tau}{2})$ and to be zero outside this interval, and considering the limit of the expression when τ goes to infinity,

$$\mathcal{F}(\sin \omega_0 t) = \lim_{\tau \to \infty} \int_{\frac{-\tau}{2}}^{\frac{\tau}{2}} \sin \omega_0 t \; e^{-j\omega t} dt$$

$$= \lim_{\tau \to \infty} \int_{\frac{-\tau}{2}}^{\frac{\tau}{2}} \frac{e^{-j(\omega-\omega_0)t}}{2j} - \frac{e^{-j(\omega+\omega_0)t}}{2j} dt$$

$$= \lim_{\tau \to \infty} \left[\frac{j\tau \sin(\omega + \omega_0)\frac{\tau}{2}}{2(\omega + \omega_0)\frac{\tau}{2}} - \frac{j\tau \sin(\omega - \omega_0)\frac{\tau}{2}}{2(\omega - \omega_0)\frac{\tau}{2}} \right]$$

$$= \lim_{\tau \to \infty} \left\{ j\frac{\tau}{2}\text{Sa}\left[\frac{(\omega + \omega_0)}{2}\right] - j\frac{\tau}{2}\text{Sa}\left[\frac{\tau(\omega + \omega_0)}{2}\right] \right\}. \quad (5.23)$$

Therefore,

$$\mathcal{F}(\sin \omega_0 t) = j\pi[\delta(\omega + \omega_0) - \delta(\omega - \omega_0)].$$

Applying a similar reasoning, it follows that

$$\mathcal{F}(\cos \omega_0 t) = \pi[\delta(\omega - \omega_0) + \delta(\omega + \omega_0)], \quad (5.24)$$

which is shown in Figure 5.13.

5.1.8 Fourier Transform of the Complex Exponential

Fourier transform can be obtained using Euler's identity,

$$e^{j\omega_0 t} = \cos \omega_0 t + j\sin \omega_0 t, \quad (5.25)$$

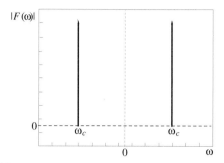

Figure 5.13 Plot for the Fourier transform of the cosine function.

and a property, as follows

$$\mathcal{F}[e^{j\omega_0 t}] = \mathcal{F}[\cos \omega_0 t + j\sin\omega_0 t]. \tag{5.26}$$

Substituting in (5.26) Fourier transforms of the sine and of the cosine functions, respectively, it follows that

$$\mathcal{F}[e^{j\omega_0 t}] = 2\pi\delta(\omega - \omega_0). \tag{5.27}$$

5.1.9 Fourier Transform of a Periodic Function

Consider the exponential Fourier series representation of a periodic function $f_T(t)$ of period T,

$$f_T(t) = \sum_{n=-\infty}^{\infty} F_n e^{jn\omega_0 t}. \tag{5.28}$$

Applying Fourier transform to both sides in Equation 5.28, it follows that

$$\mathcal{F}[f_T(t)] = \mathcal{F}\left[\sum_{n=-\infty}^{\infty} F_n e^{jn\omega_0 t}\right] \tag{5.29}$$

$$= \sum_{n=-\infty}^{\infty} F_n \mathcal{F}[e^{jn\omega_0 t}]. \tag{5.30}$$

Applying the result from (5.27) in (5.30), it follows that

$$F(\omega) = 2\pi \sum_{n=-\infty}^{\infty} F_n \delta(\omega - n\omega_0). \tag{5.31}$$

Therefore, the spectrum of a periodic function is composed of a series of Dirac's impulse functions.

6

Properties of the Fourier Transform

"Where words fail, music speaks."

Hans Christian Andersen

6.1 Properties of the Fourier Transform

Fourier transform is a powerful tool to be used in signal analysis, music, communications, control, electromagnetic theory, antenna theory, and several other areas of knowledge. It is possible to obtain a set of properties to help solve problems that require the use of Fourier transform.

6.1.1 Linearity the Fourier Transform

The application of the criteria for homogeneity and additivity makes it possible to check that the process that generates the signal

$$s(t) = A\cos(\omega_c t + \Delta m(t) + \theta), \tag{6.1}$$

from an input signal $m(t)$, is non-linear. The application of the same test to the signal

$$r(t) = m(t)\cos(\omega_c t + \theta), \tag{6.2}$$

shows that the process that generates $r(t)$ is linear.

Fourier transform is a linear operator, that is, if a function can be written as a linear combination of other functions, the corresponding Fourier transform will be given by a linear combination of the corresponding Fourier transforms of each one of the functions involved in the linear combination [Gagliardi, 1988].

If $f(t) \longleftrightarrow F(\omega)$ and $g(t) \longleftrightarrow G(\omega)$, it then follows that

$$\alpha f(t) + \beta g(t) \longleftrightarrow \alpha F(\omega) + \beta G(\omega). \tag{6.3}$$

73

Proof: Let $h(t) = \alpha f(t) + \beta g(t) \rightarrow$, then it follows that

$$H(\omega) = \int_{-\infty}^{\infty} h(t)e^{-j\omega t}dt$$

$$= \alpha \int_{-\infty}^{\infty} f(t)e^{-j\omega t}dt + \beta \int_{-\infty}^{\infty} g(t)e^{-j\omega t}dt,$$

and finally

$$H(\omega) = \alpha F(\omega) + \beta G(\omega). \tag{6.4}$$

6.1.2 Scaling Property

Fourier transform of a function that has its argument scaled by a constant a, is given by

$$\mathcal{F}[f(at)] = \int_{-\infty}^{\infty} f(at)e^{-j\omega t}dt. \tag{6.5}$$

Initially, consider $a > 0$ in (6.5). By letting $u = at$, it follows that $dt = (1/a)du$. Replacing u for at in (6.5), one obtains

$$\mathcal{F}[f(at)] = \int_{-\infty}^{\infty} \frac{f(u)}{a}e^{-j\frac{\omega}{a}u}du$$

which simplifies to

$$\mathcal{F}[f(at)] = \frac{1}{a}F\left(\frac{\omega}{a}\right).$$

Consider the case in which $a < 0$. By a similar procedure, it follows that

$$\mathcal{F}[f(at)] = -\frac{1}{a}F\left(\frac{\omega}{a}\right).$$

Therefore, finally

$$\mathcal{F}[f(at)] = \frac{1}{|a|}F\left(\frac{\omega}{a}\right). \tag{6.6}$$

This result points to the fact that if a signal is compressed in the time domain by a factor a, then its Fourier transform will expand in the frequency domain by the same factor, and vice versa.

6.1.3 Symmetry of Fourier Transform

This is an interesting property which can be fully observed in even functions. The symmetry property states that if

$$f(t) \longleftrightarrow F(\omega), \tag{6.7}$$

then it follows that

$$F(t) \longleftrightarrow 2\pi f(-\omega). \tag{6.8}$$

Proof: By definition,

$$f(t) = \frac{1}{2\pi} \int_{-\infty}^{+\infty} F(\omega)e^{j\omega t}d\omega,$$

which after multiplication of both sides by 2π becomes

$$2\pi f(t) = \int_{-\infty}^{+\infty} F(\omega)e^{j\omega t}d\omega.$$

By letting $u = -t$, it follows that

$$2\pi f(-u) = \int_{-\infty}^{+\infty} F(\omega)e^{-j\omega u}d\omega,$$

and now by making $t = \omega$, one obtains

$$2\pi f(-u) = \int_{-\infty}^{+\infty} F(t)e^{-jtu}dt.$$

Finally, by letting $u = \omega$, it follows that

$$2\pi f(-\omega) = \int_{-\infty}^{+\infty} F(t)e^{-j\omega t}dt. \tag{6.9}$$

Example: Fourier transform of a constant function can be derived using the symmetry property. Since

$$A\delta(t) \longleftrightarrow A,$$

it follows that

$$A \longleftrightarrow 2\pi A\delta(-\omega) = 2\pi A\delta(\omega).$$

6.1.4 Time Domain Shift

Given that $f(t) \longleftrightarrow F(\omega)$, it then follows that

$$f(t - t_0) \longleftrightarrow F(\omega)e^{-j\omega t_0}. \tag{6.10}$$

Let $g(t) = f(t - t_0)$. In this case, it follows that

$$G(\omega) = \mathcal{F}[g(t)] = \int_{-\infty}^{\infty} f(t - t_0)e^{-j\omega t}dt. \tag{6.11}$$

By making $\tau = t - t_0$, one obtains

$$G(\omega) = \int_{-\infty}^{\infty} f(\tau)e^{-j\omega(\tau + t_0)}d\tau \tag{6.12}$$

$$= \int_{-\infty}^{\infty} f(\tau)e^{-j\omega \tau}e^{-j\omega t_0}d\tau, \tag{6.13}$$

and finally

$$G(\omega) = e^{-j\omega t_0}F(\omega). \tag{6.14}$$

This result shows that whenever a function is shifted in time, its frequency domain amplitude spectrum remains unaltered. However, the corresponding phase spectrum experiences a rotation proportional to ωt_0.

6.1.5 Frequency Domain Shift

Given that $f(t) \longleftrightarrow F(\omega)$, it then follows that

$$f(t)e^{j\omega_0 t} \longleftrightarrow F(\omega - \omega_0). \tag{6.15}$$

$$\mathcal{F}[f(t)e^{j\omega_0 t}] = \int_{-\infty}^{\infty} f(t)e^{j\omega_0 t}e^{-j\omega t}dt \tag{6.16}$$

$$= \int_{-\infty}^{\infty} f(t)e^{-j(\omega - \omega_0)t}dt,$$

$$\mathcal{F}[f(t)e^{j\omega_0 t}] = F(\omega - \omega_0). \tag{6.17}$$

6.1.6 Differentiation in the Time Domain

Given that

$$f(t) \longleftrightarrow F(\omega), \tag{6.18}$$

it then follows that

$$\frac{df(t)}{dt} \longleftrightarrow j\omega F(\omega). \tag{6.19}$$

Proof: Consider the expression for the inverse Fourier transform

$$f(t) = \frac{1}{2\pi} \int_{-\infty}^{\infty} F(\omega)e^{j\omega t} d\omega. \tag{6.20}$$

Differentiating in time, one obtains

$$\begin{aligned}
\frac{df(t)}{dt} &= \frac{1}{2\pi} \frac{\partial}{\partial t} \int_{-\infty}^{\infty} F(\omega)e^{j\omega t} d\omega \\
&= \frac{1}{2\pi} \int_{-\infty}^{\infty} F(\omega)\frac{\partial}{\partial t} e^{j\omega t} d\omega \\
&= \frac{1}{2\pi} \int_{-\infty}^{\infty} j\omega F(\omega)e^{j\omega t} d\omega,
\end{aligned}$$

and then

$$\frac{df(t)}{dt} \longleftrightarrow j\omega F(\omega). \tag{6.21}$$

In general, it follows that

$$\frac{d^n f(t)}{dt} \longleftrightarrow (j\omega)^n f(\omega). \tag{6.22}$$

Example: By computing Fourier transform of the signal

$$f(t) = \delta(t) - \alpha e^{-\alpha t} u(t), \tag{6.23}$$

it is immediate to show that, by applying the property of differentiation in time, this signal is the time derivative of the signal

$$g(t) = e^{-\alpha t} u(t). \tag{6.24}$$

6.1.7 Integration in the Time Domain

Let $f(t)$ be a signal with zero average value, that is, let $\int_{-\infty}^{\infty} f(t)dt = 0$. By defining

$$g(t) = \int_{-\infty}^{t} f(\tau)d\tau, \tag{6.25}$$

it follows that

$$\frac{dg(t)}{dt} = f(t),$$

and since

$$g(t) \longleftrightarrow G(\omega), \tag{6.26}$$

then

$$f(t) \longleftrightarrow j\omega G(\omega),$$

and

$$G(\omega) = \frac{F(\omega)}{j\omega}. \tag{6.27}$$

In this manner, it follows that for a signal with zero average value

$$f(t) \longleftrightarrow F(\omega)$$

$$\int_{-\infty}^{t} f(\tau)d\tau \longleftrightarrow \frac{F(\omega)}{j\omega}. \tag{6.28}$$

Generalizing, for the case in which $f(t)$ has a non-zero average value, it follows that

$$\int_{-\infty}^{t} f(\tau)d\tau \longleftrightarrow \frac{F(\omega)}{j\omega} + \pi\delta(\omega)F(0). \tag{6.29}$$

6.1.8 The Convolution Theorem in the Time Domain

The convolution theorem can be used to analyze the frequency contents of a signal, to obtain many interesting results. One instance of the use of the convolution theorem, of fundamental importance in communication theory, is the sampling theorem which is discussed in the next section.

The convolution of functions $f(t)$ and $g(t)$ is defined as,

$$h(t) = f(t) * g(t) = \int_{-\infty}^{\infty} f(\tau)g(t - \tau)dt, \tag{6.30}$$

Let the resulting function have a Fourier transform $h(t) \longleftrightarrow H(\omega)$. It follows that

$$H(\omega) = \int_{-\infty}^{\infty} h(t)e^{-j\omega t}dt = \int_{-\infty}^{\infty}\int_{-\infty}^{\infty} f(\tau)g(t - \tau)e^{-j\omega t}dtd\tau. \tag{6.31}$$

$$H(\omega) = \int_{-\infty}^{\infty} f(\tau)\int_{-\infty}^{\infty} g(t - \tau)e^{-j\omega t}dtd\tau, \tag{6.32}$$

$$H(\omega) = \int_{-\infty}^{\infty} f(\tau)G(\omega)e^{-j\omega\tau}d\tau \tag{6.33}$$

and finally,

$$H(\omega) = F(\omega)G(\omega). \tag{6.34}$$

The convolution of two time functions is equivalent in the frequency domain to the product of their respective Fourier transforms.

Example: The convolution of the gate function, given by Equation 5.17, and the exponential function, given by Equation 5.10, can be used to model a sound envelope. For convenience, $a = 1$.

$$
\begin{aligned}
x(t) &= f(t) * p(t) \\
&= \int_{-\infty}^{\infty} Ae^{-(t-\tau)}u(t-\tau)[u(\tau+T/2) - u(\tau - T/2)]d\tau \\
&= \int_{-T/2}^{T/2} Ae^{-(t-\tau)}u(t-\tau)d\tau \\
&= A\left[1 - e^{-t-T/2}\right]u(t+T/2) + Ae^{-(t-T/2)}u(t-T/2).
\end{aligned}
$$

The resulting waveform is displayed in Figure 6.1, and depicts the usual attack, decay, sustain, and release parameters.

It follows from the convolution property that Fourier transform of the envelope function is the product of Fourier transforms of the exponential and gate functions,

$$X(\omega) = F(\omega) \cdot P(\omega) \tag{6.35}$$

$$= \frac{AT}{1+j\omega}\mathrm{Sa}\left(\frac{\omega T}{2}\right).$$

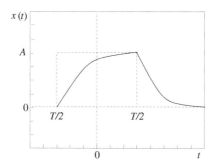

Figure 6.1   Example of an envelope signal.

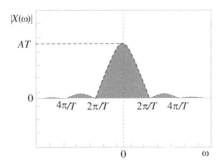

Figure 6.2   Magnitude spectrum of an envelope signal.

The spectrum magnitude of the envelope is given by the product of the corresponding magnitudes, and shown in Figure 6.2,

$$|X(\omega)| = AT\sqrt{\frac{\text{Sa}^2\left(\frac{\omega T}{2}\right)}{1+\omega^2}}. \tag{6.36}$$

Notice that the side lobes have been attenuated by the multiplication, which corresponds to a filtering of the higher frequencies. The abrupt sound, with several overtones, produced by the sharp edges of the gate function is smoothed down by the filtering process.

6.1.9 The Convolution Theorem in the Frequency Domain

For the case in which a function in time is the product of two other functions, $h(t) = f(t) \cdot g(t)$, it is possible to obtain the Fourier transform proceeding in a way similar to the previous derivation,

$$H(\omega) = \frac{1}{2\pi}[F(\omega) * G(\omega)]. \tag{6.37}$$

The product of two time functions has a Fourier transform given by the convolution of their respective Fourier transforms. The convolution operation is often used when computing the response of a linear circuit, given its impulse response and an input signal.

Example: A given circuit has the impulse response $h(t)$ as follows,

$$h(t) = \frac{1}{RC}e^{-\frac{t}{RC}}u(t),$$

in which R is a resistance, and C represents a capacitance of the circuit.

The application of the unit impulse $x(t) = \delta(t)$ as the input to this circuit causes an output $y(t) = h(t) * x(t)$. In the frequency domain, by the convolution theorem, it follows that $Y(\omega) = H(\omega)X(\omega) = H(\omega)$, that is, the Fourier transform of the impulse response of a linear system is the system transfer function.

Example: Using the frequency convolution theorem, it can be shown that

$$\cos(\omega_c t)u(t) \longleftrightarrow \frac{\pi}{2}[\delta(\omega + \omega_c) + \delta(\omega - \omega_c)] + \frac{j\omega}{\omega_c^2 - \omega^2}.$$

6.2 The Sampling Theorem

A band-limited signal $f(t)$, that has no frequency components above the maximum frequency $\omega_M = 2\pi f_M$, can be reconstructed from its samples, collected at uniform time intervals $T_S = 1/f_S$, that is, at a sampling rate f_S, in which $f_S \geq 2f_M$. In fact, the condition for uniform time intervals is not necessary.

The sampling theory, developed by Harry Nyquist (1889–1976) and Claude E. Shannon (1916–2001), has been generalized for the case of non-uniform samples, that is, samples taken at non-equally spaced intervals [Davenport and Root, 1987].

It has been demonstrated that a band-limited signal can be perfectly reconstructed from its samples, given that the average sampling rate satisfies Nyquist condition, independent of the sampling being uniform or non-uniform. [Margolis and Eldar, 2008].

For a band-limited signal $f(t) \longleftrightarrow F(\omega)$, there is a frequency ω_M above which $F(\omega) = 0$, that is, that $F(\omega) = 0$ for $|\omega| > \omega_M$. Harry Nyquist concluded that all the information about $f(t)$, shown in Figure 6.3, is

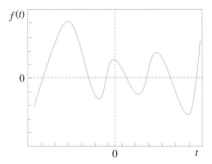

Figure 6.3   Band-limited signal $f(t)$.

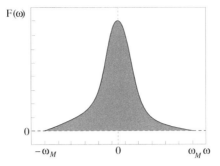

Figure 6.4   Spectrum of a band-limited signal.

contained in the samples of this signal, collected at regular time intervals T_S, as illustrated in Figure 6.4. In this way, the signal can be completely recovered from its samples.

For a band-limited signal $f(t)$, that is, such that $F(\omega) = 0$ for $|\omega| > \omega_M$, it follows that

$$f(t) * \frac{\sin(at)}{\pi t} = f(t), \text{ if } a > \omega_M,$$

because, in the frequency domain, this corresponds to the product of $F(\omega)$ by a gate function of width greater than $2\omega_M$.

The function $f(t)$ is sampled once every T_S seconds or, equivalently, sampled with a sampling frequency f_S, in which $f_S = 1/T_S \geq 2f_M$.

Consider the signal $f_s(t) = f(t)\delta_T(t)$, in which

$$\delta_T(t) = \sum_{n=-\infty}^{\infty} \delta(t - nT) \longleftrightarrow \Delta_{\omega_o}(\omega) = \omega_o \sum_{n=-\infty}^{\infty} \delta(\omega - n\omega_o). \quad (6.38)$$

The periodic signal $\delta_T(t)$ is illustrated in Figure 6.5. Fourier transform of the impulse train is represented in Figure 6.6.

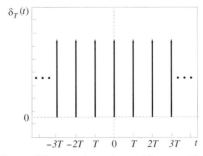

Figure 6.5   Impulse train used for sampling.

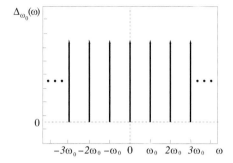

Figure 6.6 Fourier transform of the impulse train.

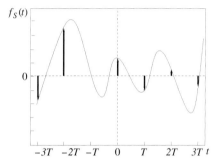

Figure 6.7 Example of a sampled signal.

The signal $f_S(t)$ represents $f(t)$ sampled at uniform time intervals T_S seconds. From the frequency convolution theorem, it follows that the Fourier transform of the product of two functions in the time domain is given by the convolution of their respective Fourier transforms. It now follows that

$$f_S(t) \longleftrightarrow \frac{1}{2\pi}[F(\omega) * \Delta_{\omega_0}(\omega)] \tag{6.39}$$

and thus

$$f_S(t) \longleftrightarrow \frac{1}{T}[F(\omega) * \Delta_{\omega_0}(\omega)] = \frac{1}{T} \sum_{n=-\infty}^{\infty} F(\omega - n\omega_o). \tag{6.40}$$

From Figures 6.7 and 6.8, it can be observed that if the sampling frequency w_S is less than $2w_M$, then the spectral components will overlap. This will cause a loss of information because the original signal can no longer be completely recovered from its samples. As the signal frequency w_S becomes smaller than $2w_M$, the sampling rate diminishes causing a partial loss of information.

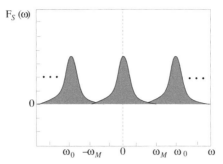

Figure 6.8   Spectrum of a sampled signal.

Therefore, the minimum sampling frequency that allows perfect recovery of the signal is $\omega_S = 2\omega_M$, and is known as the Nyquist sampling rate, after Harry Nyquist, a Swedish engineer who made important contributions to communication theory.

In order to recover the original spectrum $F(\omega)$, it is enough to pass the sampled signal through a low-pass filter with the cut-off frequency, ω_M.

Example: For the digital compact disc (CD), the typical sampling frequency is $f_S = 44.1$ k samples/s. The audio signal is then quantized, using 65,536 distinct levels. Each level corresponds to a 16-bit binary code ($2^16 = 65,536$).

After encoding, the digital music, or voice, signal is transmitted at a rate of 44.1 k samples/s $\times$ 16 bits/sample $\times$ 2 music channels $= 1411.2$ kbit/s, and the baseband signal occupies a bandwidth of approximately 1.4 MHz. This also means 10.1 Mbyte per minute of audio.

If the sampling rate f_S is lower than $2f_M$, in which f_M is the maximum signal frequency in hertz, there will be spectra overlap and, as a consequence, information loss. Therefore, the sampling frequency for a baseband signal to be recovered without loss is $f_S = 2f_M$, known as Nyquist sampling frequency.

As mentioned, if the sampling frequency is lower than Nyquist frequency, the signal will not be completely recovered, since there will be spectral superposition, leading to distortion in the highest frequencies. This phenomenon is known as aliasing, from the Latin word *alias*, meaning another, other, different.

On the other hand, increasing the sampling frequency above the Nyquist rate leads to spectra separation that is higher than the minimum necessary to recover the signal, causing a waste of spectrum usage in the transmission.

6.3 Parseval's Theorem

When dealing with a real signal $f(t)$ of finite energy, often called simply a real energy signal, the energy E associated with $f(t)$ is given by an integral in the time domain

$$E = \int_{-\infty}^{\infty} f^2(t)dt,$$

and can equivalently be calculated, in the frequency domain, by the formula

$$E = \frac{1}{2\pi} \int_{-\infty}^{\infty} |F(\omega)|^2 d.\omega$$

Equating both integrals, it follows that

$$\int_{-\infty}^{\infty} f^2(t)dt = \frac{1}{2\pi} \int_{-\infty}^{\infty} |F(\omega)|^2 d\omega. \tag{6.41}$$

The relationship given in (6.41) is known as Parseval's theorem or Parseval's identity. For a real signal $x(t)$ with energy E, it can be shown, by using Parseval's identity, that the signals $x(t)$ and its delayed version, $y(t) = x(t - \tau)$, have the same energy E.

Another way of expressing Parseval's identity is as follows.

$$\int_{-\infty}^{\infty} f(x)G(x)dx = \int_{-\infty}^{\infty} F(x)g(x)dx. \tag{6.42}$$

6.4 Average, Power, and Autocorrelation

As mentioned earlier, the average value of a real signal $x(t)$ is given by

$$\overline{x}(t) = \lim_{T \to \infty} \frac{1}{T} \int_{\frac{-T}{2}}^{\frac{T}{2}} x(t)dt. \tag{6.43}$$

The instantaneous power of $x(t)$ is given by

$$p_X(t) = x^2(t). \tag{6.44}$$

If the signal $x(t)$ exists for the whole interval $(-\infty, +\infty)$, the total average power $\overline{P}_X$ is defined for a real signal $x(t)$ as the power dissipated in a 1 ohm resistor, when a voltage $x(t)$ is applied to this resistor (or a current $x(t)$ flows through the resistor) [Lathi, 1989]. Thus,

$$\overline{P}_X = \lim_{T \to \infty} \frac{1}{T} \int_{\frac{-T}{2}}^{\frac{T}{2}} x^2(t)dt. \tag{6.45}$$

From the previous definition, the unit to measure $\overline{P}_X$ corresponds to the square of the units of the signal $x(t)$ (either volt^2 or ampère^2, depending on the use of voltage or current). These units are commonly converted to watt, using a normalization by units of impedance (ohm).

It is common use to express the power in decibels (dBm), relative to the reference power of 1 mW. The power in dBm is given by the expression [Gagliardi, 1988]

$$\overline{P}_X = 10 \log \left[\frac{\overline{P}_X}{1 \text{ mW}} \right] \text{ dBm.} \tag{6.46}$$

The total power ($\overline{P}_X$) contains two components: one constant component, because of the nonzero average value of the signal $x(t)$ ($\overline{P}_{DC}$), and an alternating component ($\overline{P}_{AC}$). The direct current (DC) power of the signal is given by

$$\overline{P}_{DC} = [\overline{x}(t)]^2. \tag{6.47}$$

Therefore, the alternating current (AC) power can be determined by removing the DC power from the total power, that is,

$$\overline{P}_{AC} = \overline{P}_X - \overline{P}_{DC}. \tag{6.48}$$

6.4.1 Time Autocorrelation of Signals

The average time autocorrelation $\overline{R}_X(\tau)$, or simply autocorrelation, of a real signal $x(t)$ is defined as follows

$$\overline{R}_X(\tau) = \lim_{T \to \infty} \frac{1}{T} \int_{\frac{-T}{2}}^{\frac{T}{2}} x(t)x(t + \tau)dt. \tag{6.49}$$

The change of variable $y = t + \tau$ allows Equation 6.49 to be written as

$$\overline{R}_X(\tau) = \lim_{T \to \infty} \frac{1}{T} \int_{\frac{-T}{2}}^{\frac{T}{2}} x(t)x(t - \tau)dt. \tag{6.50}$$

From Equations 6.49 and 6.50, it follows that $\overline{R}_X(\tau)$ is an even function of τ, and thus [Lathi, 1989]

$$\overline{R}_X(-\tau) = \overline{R}_X(\tau). \tag{6.51}$$

From the definition of autocorrelation and power, one obtains

$$\overline{P}_X = \overline{R}_X(0) \tag{6.52}$$

and

$$\overline{P}_{DC} = \overline{R}_X(\infty), \tag{6.53}$$

that is, from its autocorrelation function, it is possible to obtain information about the power of a signal. The AC power can be obtained as

$$P_{AC} = \overline{P}_X - \overline{P}_{DC} = \overline{R}_X(0) - \overline{R}_X(\infty). \tag{6.54}$$

The autocorrelation function can also be considered to obtain information about the function in the frequency domain by taking its Fourier transform, that is,

$$\mathcal{F}\{\overline{R}_X(\tau)\} = \int_{-\infty}^{+\infty} \lim_{T \to \infty} \frac{1}{T} \int_{\frac{-T}{2}}^{\frac{T}{2}} x(t)x(t+\tau)e^{-j\omega\tau} \, dt \, d\tau \tag{6.55}$$

$$= \lim_{T \to \infty} \frac{1}{T} \int_{\frac{-T}{2}}^{\frac{T}{2}} x(t) \int_{-\infty}^{+\infty} x(t+\tau) \, d\tau \, dt$$

$$= \lim_{T \to \infty} \frac{1}{T} \int_{\frac{-T}{2}}^{\frac{T}{2}} x(t)X(\omega)e^{j\omega t} \, dt$$

$$= X(\omega) \lim_{T \to \infty} \frac{1}{T} \int_{\frac{-T}{2}}^{\frac{T}{2}} x(t)e^{j\omega t} \, dt$$

$$= \lim_{T \to \infty} \frac{X(\omega)X(-\omega)}{T}$$

$$= \lim_{T \to \infty} \frac{|X(\omega)|^2}{T} \tag{6.56}$$

The power spectral density (PSD) $\overline{S}_X$ of a signal $x(t)$ is defined as the Fourier transform of the autocorrelation function $\overline{R}_X(\tau)$ of $x(t)$, that is, as

$$\overline{S}_X = \int_{-\infty}^{\infty} \overline{R}_X(\tau)e^{-j\omega\tau} \, d\tau. \tag{6.57}$$

Example: Find the PSD of the sinusoidal signal $x(t) = A\cos(\omega_0 t + \theta)$, illustrated in Figure 6.9.

Solution:

$$\overline{R}_X(\tau) = \lim_{T \to \infty} \frac{1}{T} \int_{\frac{-T}{2}}^{\frac{T}{2}} A^2 \cos(\omega_0 t + \theta) \cos\left[\omega_0(t + \tau) + \theta\right]dt$$

$$= \frac{A^2}{2} \lim_{T \to \infty} \frac{1}{T} \left[\int_{\frac{-T}{2}}^{\frac{T}{2}} \cos\omega_0\tau dt + \int_{\frac{-T}{2}}^{\frac{T}{2}} \cos\left(2\omega_0 t + \omega_0\tau + 2\theta\right) dt\right]$$

$$= \frac{A^2}{2} \cos\omega_0\tau.$$

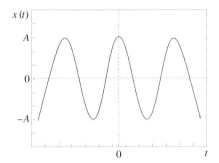

Figure 6.9   A sinusoidal signal with a deterministic phase.

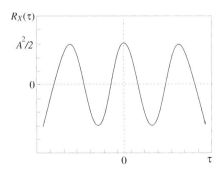

Figure 6.10   Autocorrelation function of a sinusoidal signal.

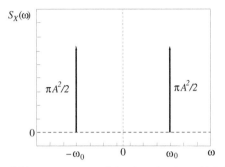

Figure 6.11   Power spectral density of a sinusoidal signal.

Notice that the autocorrelation function (Figure 6.10) is independent of the random phase θ. The PSD is given by

$$\overline{S}_X(\omega) = \mathcal{F}\left[R_X(\tau)\right]$$

$$\overline{S}_X(\omega) = \frac{\pi A^2}{2}\left[\delta(\omega + \omega_0) + \delta(\omega - \omega_0)\right],$$

which is illustrated in Figure 6.11.

The total power of the random signal $x(t)$ is computed making $\tau = 0$ in the autocorrelation function,

$$\overline{P}_X = \overline{R}_X(0) = \frac{A^2}{2}.$$

7

Musical Terminology

"Music expresses that which cannot be put into words and that which cannot remain silent."

Victor Hugo

7.1 Basic Definitions

Music is the art of combining the sounds of the human voice and of the instruments, to produce pleasure, touching the sensibility, either for joy or for sadness [Mascarenhas, 1991; Abad, 2006]. This chapter presents the usual definitions of musical terms used in music theory, such as, pitch, timbre, and beat [Palmer et al., 1994; University, 2019; Stokes, 2019].

Pitch is the relative sense of high or low of a musical sound, or note, based on frequency of vibration [Wyatt and Schroder, 1998]. Pitch is how the ear perceives the signal frequency, or the harmonics associated with it. It is possible to hear a certain pitch, even if the main frequency is absent, because the brain can still determine the pitch of the note. It is a purely psychological construct, related to the frequency of a note and to its relative position in the musical scale [Levitin, 2006].

A harmonic series is a set of vibrations produced by the main sound. The series is infinite, but the first 16 partials are sufficient to understand the sound, and also for practical applications. The timbre, the quality of a sound, produced by a musical instrument of by the human voice, is defined by the presence, or absence, and intensity of certain harmonics of the series [Chediak, 1986].

In a sense, it is like the harmonics that are heard could emulate a certain frequency that produced them. What the brain perceives as pitch is the brain's guess as to the frequency that best describe the pattern of harmonics.

Therefore, pitch can be defined as the sensation that the brain creates when interpreting repeating patterns of sound. When more than one note is

91

Figure 7.1   An example of a staff in the key of Sol (G), or treble clef.

heard at a time, the ways in which these harmonics interact can play with the brain perception of pitch in a nonsensical way [Moulton, 2019].

Harmony is the discipline that studies the best union and combination of simultaneous and distinct sounds. This study refers generally to the chords, harmonic progressions, and the structural principles that govern them [Juanilla, 2014].

It is not only the elemental category describing a vertical combinations of pitches, but also a union and combination of simultaneous and different, usually consonant, sounds. A pleasant-sounding harmony is called consonance. [Almada, 2009].

In musical notation, the *staff*, or *stave*, is composed of five horizontal lines and four spaces, and each represent a different musical pitch. In the case of a percussion staff, they represent different percussion instruments. The staff shows how the melody develops in time, that is, in music, the equivalent of the spectrogram, in signal processing. Figure 7.1 shows an example of a staff in the key of Sol (G).

While the spectrum shows the frequency amplitudes, the physical representation of the pitches, in a given moment, a spectrogram displays how frequencies change over time, as shown in Figure 7.2 [Liénard, 1977]. Appendix C presents the frequencies and wavelengths for the equal-tempered scale.

The frequencies can be represented in musical notation. A clef, from the French word *clé*, for key, is a musical symbol used to indicate the pitch of written notes. The three types of clef used in modern music notation are Fa (F), Do (C), and Sol (G). The bass clef is normally F3 (174.61 Hz), the alto clef is usually C4 (261.63 Hz), and the treble clef is G4 (392.00 Hz).

The treble clef lines, from low to high, are: Mi (E), Sol (G), Si (B), Re (D), and Fa (F); and the treble clef spaces, from low to high, are: Fa (F), La (A), Do (C), and Mi (E). The bass clef lines, from low to high, are: Sol (G), Si (B), Re (D), Fa (F), and La (A); and the bass clef spaces, from low to high,

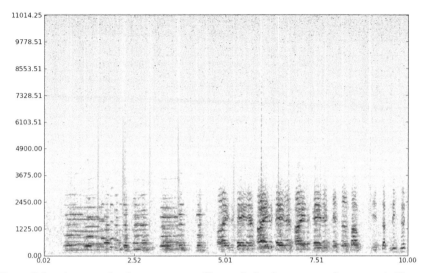

Figure 7.2 An example of a spectrogram. The vertical axis represents frequency, in Hz, and the horizontal axis has time, in seconds. (Adapted from http://www.frank-zalkow.de/en/code-snippets/ create-audio- spectrograms-with- python.html?i=1. Licensed under a Creative Commons Attribution 3.0 Unported License. Frank Zalkow, 2012–2013.)

are: La (A), Do (C), Mi (E), and Sol (G). If a pitch extends beyond the range of a clef, it is indicated by a short line called *ledger line* [Wyatt and Schroder, 1998].

An interval is the measured distance between two musical pitches, or two notes in a diatonic scale, such as the major second, a whole step, Do to Re (C-D) interval, or the semitone, or half-step, interval Mi to Fa (E-F). A compound interval is an interval larger than an octave.

The interval is the building block of the chord. It can be melodic, horizontal, or linear, if it relates to tones that sound in sequence, or notes that are played one after another. The harmonic, or vertical, interval implies tones that sound simultaneously, or notes that are simultaneously played, such as in a chord.

The intervals can be either ascending or descending, consonant or dissonant. A simple interval spans one octave, and a compound interval spans more that one octave, but usually not more than two octaves, as presented in Table 7.1 [Powell, 2010; Almada, 2009].

The tone has a specific duration. A musical measure is the section of a musical staff that appears between two barlines. A common term for a musical measure is the *bar*. A bar, or measure, is a group of beats, or pulses.

Table 7.1 Classification of intervals

No. semitones	Names	Symbols	Classification
0	Perfect unison	P1 or 1	Diminished second
1	Minor second	m2	Augmented unison
2	Major second	M2	Diminished third
3	Minor third	m3	Augmented second
4	Major third	M3	Diminished fourth
5	Perfect fourth	P4	Augmented third
6	Tritone	A4, 5° or d5	Dim. fifth, Aug. fourth
7	Perfect fifth	P5	Diminished sixth
8	Minor sixth	m6 or 5+	Augmented fifth
9	Major sixth	M6	Diminished seventh
10	Minor seventh	m7 or 7	Augmented sixth
11	Major seventh	M7	Diminished octave
12	Perfect octave	P8 or 8	Augmented seventh
13	Minor second	m9	Minor ninth
14	Major second	M9	Major ninth
15	Minor third	m10	Minor tenth
16	Major third	M10	Major tenth
17	Perfect fourth	11th	Eleventh

The most common grouping of beats is four in a bar, called common time [Hammer, 1999].

The end of a musical part is referred to as the *coda*. The *coda* is a concluding section appended to the end of a work. It means tail in Italian. *Da capo* is an Italian expression that means to the head, that is, a written indication telling a performer to go back to the start of a piece, and *da capo al fine* means from the beginning to the end [Friedland, 2004].

A *ternary form* is the usual ABA scheme, that indicates statement, contrast, and restatement. A transition is a bridge section between two musical ideas. The transposition shifts a piece to a different pitch level.

A *tremolo* is a rapid repetition of a pitch, that is, bowing a string rapidly while maintaining a constant pitch. On the other hand, a *trill* is a rapid alternation of two close pitches to create a shaking ornament on a melodic note.

7.2 A Measure of Dynamics in Music

Dynamics is the musical element of relative musical loudness or quietness. Italian terms are commonly used to indicate dynamics [Shepheard, 2014].

Pianissimo is a very quiet, or very soft, dynamic marking (*pp*). *Piano* is a quiet, or soft, dynamic marking (*p*). *Mezzo-piano* is a medium quiet dynamic marking (*f*).

Mezzo-forte – A medium loud dynamic marking (*F*). The Italian term *forte* indicates a loud dynamic marking (*f*), while *fortissimo* is a very loud dynamic marking (*ff*).

A change in dynamics is also indicated by Italian words. For example, *sforzando* means a sudden stress on a note or chord (*sfz*), while *smorzando*, or *morendo*, indicates that the sound is dying away (*smz*). *Calando* means gradually softer and slower.

When the sound intensity gradually gets louder it is called a *crescendo*. Of course, *decrescendo*, or *diminuendo*, means that the sound intensity is gradually getting quieter.

7.3 A Measure of Time in Music

The *metronome* is a mechanical, electric, or electronic device that precisely measures tempo, and is used by musicians to synchronize the instruments. The beat is a musical pulse, the basic unit of time, but, usually, it can refer to a variety of related concepts, which include: pulse, tempo, meter, and specific rhythms.

A very slow, broad tempo, is called *largo*, which corresponds to 40–60 beats per minute, while *larghetto* is a slow tempo, not quite as slow as *largo*, and corresponds to 60–66 beats per minute [Abad, 2006].

The Italian term *adagio* means a slow tempo, while *allegro* is a fast tempo, which corresponds to 66–76 beats per minute. *Andante* is faster than *adagio*. Also implying a walking speed, because *andare* means to walk, in Italian. It corresponds to 76–108 beats per minute [Abad, 2006].

Moderato is a moderate tempo, which corresponds to 108–120 beats per minute, while the Italian term *allegro* means a brisk and lively movement, that corresponds to 120–168 beats per minute [Abad, 2006].

Presto is a very fast tempo, which corresponds to 168–200 beats per minute, and *prestissimo* means as fast as possible, which corresponds to 200–208 beats per minute [Abad, 2006].

Vivace is a lively tempo, while a slow, solemn tempo, is called *grave* [Cebolo, 2015]. On the other hand, *rubato* is a flexible approach to metered rhythm, in which the tempo can be momentarily sped up, or slowed down, to improve the personal expression of the performer.

Ritardando means slowing down the tempo, while the term *accelerando* means gradually accelerating the speed of the rhythmic beat. Augmentation means lengthening the rhythmic values of a fugal subject. *Affrettando* indicates that the musician must hurry up, and *incalzando* means to follow closely [Cebolo, 2015].

The *downbeat* is the first beat of a musical measure, usually accented more strongly than other beats. An accent is an instantaneous emphasis in a note, using a dynamic attack. The *duple meter* is a basic metrical pattern having two beats per measure. The *triple meter* is a common meter with three beats per measure.

7.4 Musical Notation and Accidentals

An accidental is a note of a pitch, or pitch class, that is not a member of the scale or mode indicated by the most recently applied key signature. In musical notation, the sharp (♯), flat (♭), and natural (♮) symbols are the basic symbols to mark such notes, and are also called accidentals [Bennett, 1998].

As discussed earlier, the bemol is a musical symbol that lowers the pitch by one half-step, or by one semitone. It is represented by the symbol flat sign (♭), a stylistic abbreviation of the word *bemol*. The sharp sign is a musical symbol that raises the pitch one half-step.

Tie – A curved line connecting two notes, an indication to hold the tone for the combined rhythmic value of both notes.

Slash chord – Chord symbol indicating a specific bass note to play.

Beam – A horizontal line that connects and replaces a group of short notes.

A tempo – It is a return to the original tempo, that usually appears after a *accelerando* ou *ritardando* [Neely, 1988].

Bis – An instruction to repeat a short passage.

Crotchet – A note worth one beat within a bar of four-four tone.

Mediant – The third degree of the major scale.

Mordent – An ornamental instruction to play a single note as a trill with an adjacent note.

Perfect cadence – A cadence that resolves from the dominant to the tonic.

Ad libitum – Instruction that the performer may freely interpret or improvise a passage of a music piece [Burrows, 2002].

Expression marks – Words or symbols written on a score to guide the player on dynamics, articulation, and tempo.

Dal segno **(DS)** – Means from the sign, indicating that the performer must repeat a sequence from a point marked by a symbol [Burrows, 2004].

Cadence – A melodic or harmonic relevant configuration that produces a sense of resolution, and usually closes a musical composition, or any part of it.

Legato – A smooth, connected manner of performing a melody.

Libretto – The sung or spoken text of an opera.

Film music is a music genre that serves either as background or foreground material for a movie.

Glissando means a rapid slide between two distant pitches.

The half-step is the smallest interval in the Western system of equal temperament.

Mazurka – A type of Polish dance in triple meter, sometimes used by Chopin in his piano works.

Medieval – A term used to describe things related to the Middle Ages (c.450–1450).

Melisma – A succession of many pitches sung while sustaining one syllable of text.

Mezzo – An Italian prefix that means medium.

Microtone – A non-Western musical interval that is smaller than a Western half-step.

MIDI – An acronym for musical instrument digital interface. It is a protocol established in the 1970s that allows digital synthesizers to communicate with computers.

Minimalism – A modern compositional approach promoted by Glass, Reich, and other composers, in which a short melodic, rhythmic, or harmonic idea is repeated and gradually transformed as the basis of an extended work.

Minuet – An aristocratic dance in 3/4 meter.

Minuet and trio form – The traditional third-movement form of the Classic four-movement design, based on an aristocratic dance in 3/4 meter.

Motet – A polyphonic vocal piece set to a sacred Latin text that is not from the Roman Catholic mass.

Motive – A small musical fragment used to build a larger musical idea. It can be reworked in the course of a composition, as in the four-note motive in Beethoven's Symphony No. 5, in C minor.

Movement – A complete, independent division of a larger work.

Mp3 – A modern technology that allows digital sound to be compressed into files that are approximately eight times smaller than the original, with small loss of quality.

Mute – A device used to muffle the tone and volume of an instrument.

Natural sign – A musical symbol that raises the pitch one half-step.

Notation – A system for writing music down so that critical aspects of its performance can be recreated accurately.

Orchestra – A large instrumental ensemble comprised of strings, wood-winds, brasses, and percussion.

Orchestration – The technique of conceiving or arranging a composition for orchestra.

Ostinato – A short rhythmic or melodic idea that is repeated exactly over and over throughout a musical section or work.

Percussion instrument – An instrument on which sound is generated by striking its surface with an object.

Pizzicato – Refers to a type of violin playing in which a string is plucked by the fingers.

Polka – A lively Bohemian (Czech) dance, traditionally for the common classes.

Polonaise – A Polish nationalistic military dance used in some of Chopin's piano character pieces.

Postlude – A concluding section, usually at the end of a keyboard movement.

Progression – A series of chords that functions similarly to a sentence or phrase in written language.

Quadruple meter – A basic metrical pattern having four beats per measure.

Quotation music – A composition extensively using quotations from earlier works, common since c.1960.

Raga – A melodic pattern used in the music of India.

Recapitulation – The third aspect of Classic sonata form. In this section, both themes of the exposition are presented in the tonic key.

Restated – In the home key. The second theme gives up its opposing key center.

Recitative – A speech-like style of singing with a free rhythm over a sparse accompaniment.

Recorder – An ancient wooden flute.

Register – A specific coloristic portion of an instrumental or vocal range.

Riff – The repetitive series of notes, chord pattern, or musical phrase motive. It is frequently used as an introduction to a song, such as, a guitar riff [Moon, 2015].

Ritornello form – A Baroque design that alternates big and small effects, as in *tutti versus solo*. Usually, the *tutti* section is a recurring melodic refrain.

Score – Written notation that vertically aligns all instrumental/vocal parts used in a composition.

Singspiel – A traditionally low-level type of comic light opera, featuring spoken German dialog interspersed with simple German songs.

Sitar – A long-necked stringed instrument of India.

Sprechstimme – A half-spoken, half-sung style of singing on approximate pitches, developed by Schönberg in the early 1900s.

Staccato – Short, detached notes.

Strophic form – A song form featuring several successive verses of text sung to the same music.

Subject – The main melodic idea of a fugue.

Suite – A collection of dance movements.

Synthesizer – A modern electronic keyboard instrument capable of generating a multitude of sounds.

Tala – A rhythmic pattern used in the music of India.

Tone cluster – A modern technique of extreme harmonic dissonance created by a large block of pitches sounding simultaneously.

Tone row – An ordered series of 12 chromatic pitches used in serialism.

Trio sonata – A Baroque multi-movement chamber work for four performers, including two violins and *basso continuo*.

Triplet – A rhythmic grouping of three equal-valued notes played in the space of two, that is indicated in written music by the symbol "3" above the grouping.

Tutti – Italian word for all or everyone. It is an indication for all performers to play together.

Unison – The rendering of a single melodic line by several performers simultaneously.

Upbeat – The weak beat that comes before the strong downbeat of a musical measure.

Variation – The compositional process of changing some aspects of a musical work, while retaining others.

Verismo – A style of true-to-life Italian opera that flourished at the turn of the 20th century.

Vibrato – Small fluctuations in pitch used to make a sound more expressive.

Virtuoso – A performer of extraordinary ability.

Volume – The relative quietness or loudness of an electrical impulse.

Waltz – An aristocratic ballroom dance in triple meter that flourished in the Romantic period.

Whole step – An interval twice as large as a half-step. Example: the distance between C and D on a piano.

Word painting – In vocal music, musical gestures that reflect the specific meaning of words. It is a common aspect of the Renaissance madrigal.

World beat – The collective term for contemporary popular third-world musical styles. Also called ethno pop.

Gamelan – An Indonesian musical ensemble comprised primarily of percussion instruments.

Ensemble is a group of musical performers.

Étude – A French word for a study piece, designed to help a performer master a particular technique. The form is the elemental category that describes the shape or design of a musical work or movement.

Tablature (TAB) – A simplified system of notation used for the guitar, that dates back to the Renaissance period.

Back beat – A term used to describe the emphasis of the weak beats two and four, in common time (4/4) [Capone, 2007].

Barre – The method of placing the first finger, or other fingers, across adjacent strings to hold down adjacent chord notes [Chapman, 2003].

Improvisation – An instantaneous creation of music, while it is being performed. It is usual in American jazz and rap; but also in Brazilian *embolada* and *funk*. Usually, improvisation requires a greater knowledge of music theory [Chediak, 1986].

8

Sound Theory

"Music produces a kind of pleasure which human nature cannot do without."

Confucius

8.1 Sound Characteristics

Sound is a vibration that can propagate as an audible wave of pressure, through a material transmission medium such as a gas, liquid, or solid. It is the sensation produced in the ear by the vibratory movement of the bodies, transmitted by an elastic medium [Abad, 2006].

From a psychological point of view, the sound is the reception of such waves and their perception by the brain. The movement of the oscillations in a sound wave is longitudinal, that is, parallel to the direction of movement overall [Tan et al., 2018].

Human beings can only hear sound waves as distinct pitches when the frequency lies between 20 Hz and 20 kHz. Sound waves above 20 kHz are known as ultrasound, and below 20 Hz are called infrasound.

The period T of a sound wave is the reciprocal of its frequency,

$$T = \frac{1}{f} = \frac{\lambda}{c} \text{ s,} \tag{8.1}$$

in which, λ is the wavelength, given in meters, c is the sound velocity, in m/s, and the frequency f is given in Hz.

From Formula 3.17,

$$\omega = k\sqrt{\frac{U}{m/l}},$$

and recalling that $\omega = 2\pi f$ and $k = 2\pi/\lambda$, one obtains

$$2\pi f = \frac{2\pi}{\lambda} \sqrt{\frac{U}{m/l}}, \tag{8.2}$$

which, after simplifying, gives

$$f = \frac{1}{\lambda} \sqrt{\frac{U}{m/l}}. \tag{8.3}$$

But, the condition for the fundamental standing wave is $\lambda = 2l$, therefore the fundamental frequency of a tensioned string is

$$f = \frac{1}{2l} \sqrt{\frac{U}{\mu}} \;\; \text{Hz}, \tag{8.4}$$

in which, $l = \lambda/2$ is the string length, in meters, m is the string mass, in grams, and U is the string tension, in newtons. The linear density, that is, mass per unit length of the string, is $\mu = m/l$.

8.1.1 The Sounds of an Acoustic Guitar

Note from Formula 8.4 that the sound frequency, or pitch, is inversely proportional to the string length, which every musician's experience confirms. It is also directly proportional to the square root of the string tension, which is expected, and inversely proportional to the square root of the mass density of the string, as can be verified by weighting the strings of a guitar, for example.

Therefore, to create a sound of a higher pitch on a tuned acoustic guitar, it is necessary to pinch a string to shorten it, as indicated by Formula 8.4. For instance, if the player strums the sixth string (chromatic note Mi), which corresponds to E2 in the scientific pitch notation, the frequency is $f = 82.41$ Hz. This results is obtained for the standard tuning (International System of Units – SI), in which the chromatic note La (A4) frequency is $f = 440$ Hz.

On the other hand, if the player pinches the same string in the middle, the 12th fretboard, the frequency doubles to $f = 164.82$ Hz, which is also Mi, and corresponds to E3. This is one octave above the original note. Figure 8.1 shows how to use an acoustic guitar to produce a sound of one octave above the fundamental. The player has to pinch the string exactly in the middle.

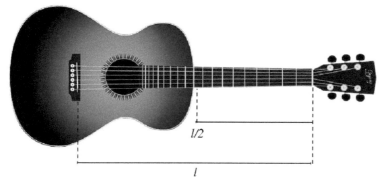

Figure 8.1 Using an acoustic guitar to produce a sound one octave above the fundamental pitch.

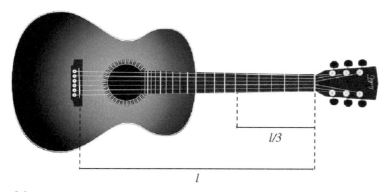

Figure 8.2 Using an acoustic guitar to produce a sound one-fifth above the fundamental pitch.

If the player pinches the same string in a fretboard that is one-third of the string length from the nut, that leaves 2/3 of the string to be played, the obtained chromatic note is Si, the frequency is $f = 123.62$ Hz, which corresponds to B2. This is one-fifth above, or 3/2 times the pitch of the original note, and can be used to tune the fifth string. Figure 8.2 shows how to use an acoustic guitar to produce a sound one-fifth above the fundamental.

If the player pinches the same string in the fifth fretboard that is one-fourth of the string length from the nut, that leaves 3/4 of the string to be played, the obtained note is La, the frequency is $f = 110$ Hz, which corresponds to A2. This is one-fourth above, or 4/3 times the pitch of the original note, and can be used to tune the fifth string. Figure 8.3 shows how to use an acoustic guitar to produce a sound one-fourth above the fundamental.

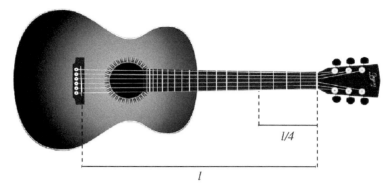

Figure 8.3   Using an acoustic guitar to produce a sound one fourth above the fundamental pitch.

At 20° C, the velocity of sound in air is 343 m/s, or 1,234.8 km/h. The velocity of sound in air is approximated by the formula, [Ashton, 2001]

$$v = 331.5 + 0.60Q \ \text{m/s}, \tag{8.5}$$

in which Q is the air temperature, in Celsius, and the velocity is in m/s. Therefore, for example, the velocity of sound through air is 346 m/s at 25° C.

An overtone is any frequency above the fundamental frequency of a sound. The fundamental and the overtones together are called partials. Harmonics partials have frequencies that are numerical integer multiples of the fundamental. The overtones define the *timbre* of an instrument, which depends on the partials that are emphasized by the instrument's structure.

Figure 8.4 illustrates harmonic overtones, or partials, on a stationary wave along a string when it is held fixed at certain lengths, as when a guitar string is plucked while held half way along its length, on the 12th fretboard.

Sounds that present a rough timbre are composed of overtones that are not simply related to the fundamental, and they are called anharmonic complex tones. Many percussion instruments, such as, drum, bass drum, tambourine, maracas, gongs, chimes, and bell plates include harmonics that do not conform to integer relationships [Tan et al., 2018].

8.1.2 The Acoustic Guitar Characteristics

A guitar is constructed from several parts, including the soundboard, the tuning knobs, the clamps, the sides and faces of the air chamber and, typically, six strings. The strings produce six fundamental approximate

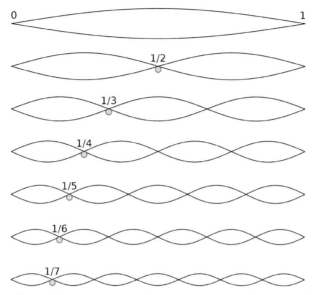

Figure 8.4 Harmonic overtones on the wave set up along a string. (Adapted from https://en.wikipedia.org/wiki/Overtone#/media/File:Harmonic_partials _on_strings.svg. Public domain.)

frequencies: 82 Hz, 110 Hz, 147 Hz, 196 Hz, 247 Hz, and 330 Hz corresponding to the open standard notes of E2, A2, D3, G3, B3, and E4.

The guitar is a mechanical system, with a cavity, that has its dynamic behavior determined by the interaction of several components [Curtu et al., 2009]. The plucked strings radiate only a small amount of sound energy directly, but they excite the bridge and the soundboard, which in turn transfer energy to the air cavity, ribs, and resonance box. Sound is radiated efficiently by the vibrating parts and through the sound hole [Elejabarrieta et al., 2002].

The resonant frequency of a cavity with an opening is [Ashton, 2001]

$$f = \frac{c}{2\pi}\sqrt{\frac{A}{Vl}} \text{ Hz}, \tag{8.6}$$

in which, l is the opening length, c is the sound velocity, A is the opening area, and V is the volume of cavity.

The main components of the guitar body that produce sound are the soundboard, which resonates more with high frequency pitches, and the resonance box, which has a tendency to vibrate with lower frequency pitches.

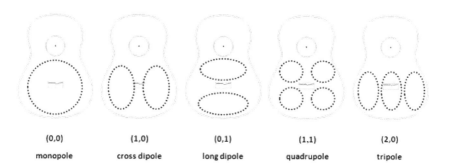

| (0,0) | (1,0) | (0,1) | (1,1) | (2,0) |
| monopole | cross dipole | long dipole | quadrupole | tripole |

Figure 8.5 Primary modes of vibration for the guitar top. (Adapted from Eendebak, 2019.)

With the sound being transferred into the wood from the strings via the bridge, a resonance pattern is formed on the soundboard of the guitar.

Figure 8.5 shows the modes of vibration for a traditional classical guitar top. Some fundamental frequencies below 500–600 Hz are most important, and generate the (0.0), (1.0), (0.1), (1.1) and (2.0) modes as shown in the figure [Eendebak, 2019].

The natural frequencies of the guitar resonance box are obtained with the fixed ribs from experimental measurements, analyzed by the modal analysis technique. The back is viewed as if the soundboard were transparent [Elejabarrieta et al., 2002].

8.1.3 The Plucking Effect

As every musician have noticed, the position a string is plucked makes a difference, regarding the instrument timbre, or the production of sound overtones. For instance, the harmonics produced by plucking the string in the middle, near the 12th fret, are different from those produced by plucking the string very close to the guitar bridge.

This is the result of the solution of the wave Equation 3.7, that is repeated in the following for convenience, under certain initial, or boundary, conditions.

$$\frac{\partial^2 u}{\partial x^2} = \frac{1}{c^2}\frac{\partial^2 u}{\partial t^2}. \tag{8.7}$$

The general solution to the wave equation, for a vibrating string of length l with fixed ends, can be written as the sum of sinusoids representing the normal modes [Traube and Smith, 2000],

$$u(x,t) = \sum_{n} \left[a_n \sin(\omega_n t) + \cos(\omega_n t) \right] \sin(k_n x), \tag{8.8}$$

in which

$$a_n = \frac{2}{l\omega_n} \int_0^t u'(x,0) \sin\left(\frac{n\pi x}{l}\right) dx, \tag{8.9}$$

and

$$b_n = \frac{2}{l} \int_0^t u(x,0) \sin\left(\frac{n\pi x}{l}\right) dx. \tag{8.10}$$

Therefore, the amplitude of the nth mode is given by

$$c_n = \sqrt{a_n + b_n}. \tag{8.11}$$

Consider an ideal plucking excitation, at a distance r from an end, for example the guitar bridge, and with an amplitude A. The initial velocity of the points on the string, for any x, is

$$u'(x,0) = 0, \tag{8.12}$$

and the string is initially shaped as a triangle, with the peak at the point (r, A). Then,

$$u(x,0) = \frac{Ax}{r}, \quad 0 \le x \le r, \tag{8.13}$$

and

$$u(x,0) = \frac{A(l-x)}{l-r}, \quad r \le x \le l. \tag{8.14}$$

For those conditions, $c_n = b_n$, and the solution of the Integral 8.10 gives

$$c_n = \frac{2A}{n^2\pi^2 q(1-q)} \sin(n\pi q), \tag{8.15}$$

in which $q = r/l$ is the fraction of the string length from the point where the string is plucked to the bridge.

For example, when $u = u(x,0)$, the vertical displacement of a point on the string at time zero, is set as an isosceles triangle, as shown in Figures 8.6 and 8.7, only odd modes are excited in the string. Besides that, the magnitude of the even modes are inversely proportional to the square of the mode number [Loy, 2011a].

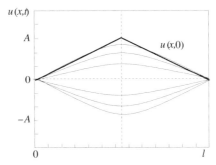

Figure 8.6   A plucked guitar string, and the excited modes.

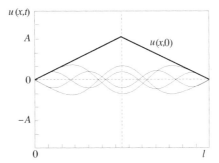

Figure 8.7   A plucked guitar string, and some more excited modes.

The even nodes are not excited, and the sound spectrum has a concentration of energy in the first mode, corresponding to the fundamental frequency, or the basic pitch.

When the plucking position is near the guitar bridge, the number of harmonics increases, and their amplitude decrease only with the inverse proportion of the coefficient number. Therefore, plucking closer to the bridge, the end of the string, causes higher-order modes to receive more energy, resulting in different timbre, and a brighter sound.

This effect is illustrated in Figure 8.8, in which one observes that the overtones produced by plucking the string in the middle, the solid lines, decrease rapidly. The string plucked near the bridge produces more harmonics, the dashed lines, with more energy in higher frequencies.

Observe that the eighth overtone is missing, because of the initial constraints of the wave equation solution. These are spatial sinusoids, as a function of position, but they also correspond to natural frequencies of vibration of the string.

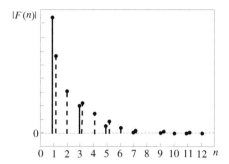

Figure 8.8 The spectrum produced by plucking a guitar string.

The points that do not displace from their position are called nodes, and correspond to the zero crossings of the sinusoidal functions. The end points are forced nodes, because they are fixed, for all possible excitation conditions. If a string is plucked in the middle, for instance, only waveforms that do not present nodes in the middle can appear, that is, only the odd-numbered modes are allowed. The even modes do not contribute to the triangular shape of the string, one of the boundary conditions [Cook, 2003].

8.2 Sound Envelope

Each musical instrument is characterized by a different envelope, that is a combination of intensities through which the sound manifests itself, since its inception till its end. There are four main parameters that distinguish the envelope, that is, attack, decay, sustain, and release, known by the acronym ADSR, that are illustrated in Figure 8.9 [Abad, 2006].

Attack (A) – It is the interval during which the instrument begins to sound, using pressure, strumming, or blow. It is the time that it takes for a signal to reach the highest point of amplitude after being triggered [Henrique, 2014].

Decay (D) – It is the posterior interval, when the musical instrument begins to loose energy. Decay is the time it takes to drop down to the sustain level, after reaching the initial peak of the attack period.

Sustain (S) – It is the interval during which the pressure or action on the material or column of air that are set to vibrate is maintained. It is also a

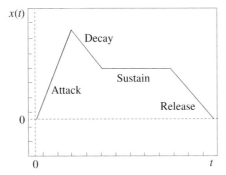

Figure 8.9   General form of the envelope of a sound.

level of amplitude that the signal remains on for as long as the sound is being triggered.

Release (R) – This is the moment during which the action of the material or air column ceases. The release parameter determines how long it takes for the sound to fade out completely from the sustain level.

When observed on the screen of an oscilloscope, the modulated signal

$$s(t) = x(t)\cos(\omega t + \phi), \qquad (8.16)$$

a mathematical model of the sound produced by an instrument or by a synthesizer, looks like the one depicted in Figure 8.10. Recall that $\omega = 2\pi f$, and f is the fundamental pitch of the instrument, in Hz. The associated intervals are also shown in the figure.

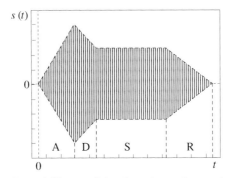

Figure 8.10   Modulated envelope of a sound.

The sound becomes music as it takes on different durations, that produce the rhythmic figures; as it assumes diverse pitches, that generate melodic and harmonic movements; as it displays particular timbres, that multiply the color of the voices; and as it shows several intensities, the curves and folds that give strength to the sound emission.

9

Acoustics and Cognition

"Great music is that which penetrates the ear with facility and leaves the memory with difficulty. Magical music never leaves the memory."

Sir Thomas Beecham

9.1 Acoustics and the Brain

The brain hemispheres are specialized in certain auditory activities, which probably evolved to keep the areas that are responsible for the input and processing of speech in close proximity with the gestural, nonverbal, and vocal areas, in order to minimize the transmission delays between the acting neural nets. Table 9.1 presents a comparative list of auditory characteristics related to the left and right brain hemispheres [Roederer, 1998].

The complex sequential operations involved in speech processing do not allow enough time for the communication between the two hemispheres, which require approximately 50 ms. Therefore, the interpretative, integrative, and holistic functions, that are time consuming, were taken by the right hemisphere. The brain recognizes the musical messages as having a holistic nature, presenting patterns of longer duration, as opposed to short sequences.

The music is identified by the brain as a representation of integral holistic auditory images, that is, a harmonic structure. The brain perceives the music as a temporal sequence of long duration, with a certain melodic contour. The left temporal lobe is specialized on verbal inputs, and the music inputs are processed by the right hemisphere [Roederer, 1998].

9.2 The Unit of Loudness

In acoustics, sound pressure level (SPL), or acoustic pressure level, is a logarithmic measure of the effective pressure of a sound relative to a reference

Table 9.1   A comparative list of auditory characteristics

Left hemisphere	Right hemisphere
Plosive consonants	Continuous vowels
Syntax and phonological attributes	Stereotyped attributes, poetic rhymes
Speech understanding	Voice intonation, ambient and animal sounds
Propositional speech	Emotional content of speech
Analysis of nonsense spoken sounds	Pitch, timbre, tonality and harmony
Spoken text, verbal content	Sung text, musical and phonetical content
Rhythm, short sequences of melodic sounds	Holistic melody
Verbal memory	Tonal memory

value, usually $P_0 = 20 \ \mu\text{Pa}$ at a frequency of 1 kHz, which is considered as the threshold of human hearing [Benade, 1990].

The density of the sound pressure is given by

$$I = \frac{P}{A}, \tag{9.1}$$

in which, P is the sound pressure, and A is the area.

For an ideal point source, the spherical sound wave, that represents the intensity in the radial direction as a function of distance r from the center of the sphere, is given by

$$I = \frac{P}{4\pi r^2}. \tag{9.2}$$

The lower limit for the perceived sound pressure is 1×10^{-12} W/m^2, in SI units.

The sound pressure level, denoted L_p and measured in dB, is defined by

$$L_P = 20 \log_{10}\left(\frac{P}{P_0}\right) \text{dB}, \tag{9.3}$$

in which, P is the root mean square sound pressure and P_0 is the reference sound pressure. The lower limit of audibility is defined as $L_p = 0$ dB.

Those are physical units, that can be measured by a standard equipment. But, the perceived sensation, which involves the auditory system and the brain, requires a different approach.

Table 9.2 A comparison between levels of sones, phons and musical notation

Sone	1	2	4	8	16	32	64	128
Phon	40	50	60	70	80	90	100	110
Dynamic	*ppp*	*pp*	*p*	*mp, mf*	*f*	*ff*	*fff*	*ffff*

There are two usual units to access the sound perception by the human ear, the sone and the phon. The sone is a unit of loudness, and indicates how loud a sound is perceived by the auditory system. Doubling the perceived loudness doubles the sone value. It is not an SI unit [Davis and Davis, 1987].

The phon is a unit of loudness level for pure tones. This implies that zero phon is the limit of perception. The phon is also used to indicate an individual's perception of loudness. By definition, 1 phon is equivalent to 1 dB at 1000 Hz (1 kHz).

The purpose of the phon is to provide a standard measurement for perceived intensity. The phon unit is not an SI unit in metrology, but it is used as a unit of loudness level by the American National Standards Institute (ANSI).

According to the sone scale, 1 sone sound is defined as a sound whose loudness is equal to 40 phons, in other words, a loudness of 1 sone is equivalent to the loudness of a signal at 40 phons, and this represents the loudness level of a 1 kHz tone at 40 dB SPL.

Table 9.2 shows a comparison between levels of sones, phons, and traditional musical dynamics notation.

At frequencies different from 1 kHz, the loudness level in phons is adjusted according to the frequency response of the human hearing, using a set of equal-loudness contours, and then the loudness level in phons is mapped to loudness in sones using the same power law. Loudness S in sones (for $L_P > 40$ phon) [Davis and Davis, 1987],

$$S = 2^{\frac{L_P - 40}{10}}. \tag{9.4}$$

The loudness level L_P in phons (for $S > 1$ sone),

$$L_P = 40 + 10 \log_2(S). \tag{9.5}$$

The formulas are valid for specific values, and corrections are needed at lower levels, for instance, near the threshold of hearing. Figure 9.1 shows equal-loudness contours for human hearing. The dark line represents a contour for 40 phons.

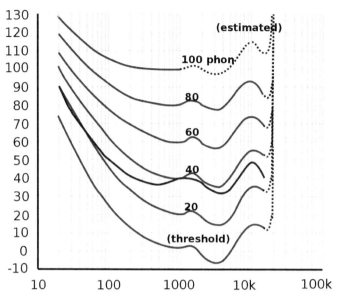

Figure 9.1   Equal-loudness contours, from ISO 226:2003 revision, with frequency in Hz, and loudness measured in dB SPL. (This image is adapted from Wikimedia Commons freely licensed media file repository.)

The loudness of a sinusoid, or narrow-band noise, is also related to its pressure amplitude according to the equations that follow [Benade, 1990],

$$S = \left(\frac{P}{100P_0} \right)^{0.6}, \tag{9.6}$$

in which, S is given in sones. This formula is valid for a signal whose pressure amplitude P is 40 dB, one 100 times above the threshold value, P_0.

On the other hand, for a weak signal, in the immediate neighborhood of the threshold, the formula becomes,

$$S = \left(\frac{P}{22P_0} \right)^{2}. \tag{9.7}$$

As the signal level increases in sound pressure from the threshold, the factor (22) in the denominator grows smoothly toward 100, and the exponent decreases from 2 to 0.6, a factor that applies for the sound pressures above that are 40 dB above the threshold. It is interesting to note that this effect of sound compression at higher levels is common for every human sense. It is probably a security measure of the body to protect the sensory system.

A distorted electronic system sounds louder than an undistorted one at equal power levels. For example, if two equal level signals are produced near 1000 Hz, with minimum spectral separation, and are gradually separated, while maintaining their phon level, they will seem to increase in apparent loudness.

9.3 The Auditory System

The auditory system is composed of the external ear, that receives the sound waves and directs them to the auditory canal; the middle ear, that includes the ossicles; and the inner ear, whose main organ is the cochlea, a complex set of sound detectors to transduce certain characteristics of the sound into neural impulses to be transmitted to different parts of the brain [Tan et al., 2018].

The cochlea is an organ of the inner ear that contains the sensory organ of hearing. The cochlea is a spiral tube that is coiled around a hollow central pillar, the modiolus. It is a conical chamber of bone, in which waves propagate from the base, near the middle ear and the oval window, to the apex, the top or center of the spiral [Joseph E. Hawkins, 2019]. Figure 9.2

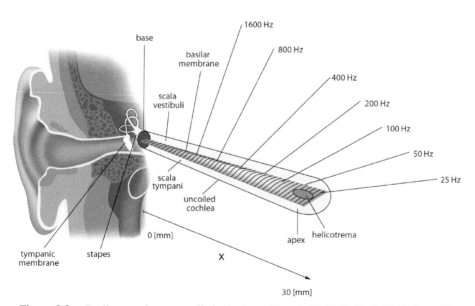

Figure 9.2 Basilar membrane uncoiled. (Authors: Kern A, Heid C, Steeb W-H, Stoop N, Stoop R. – This image is licensed under the Creative Commons Attribution-Share Alike 2.5 Generic license.)

shows an uncoiled membrane, indicating the points of resonance of particular frequencies. It functions as a spectrum analyzer.

The cochlea is filled with a liquid, called endolymph, that moves in response to the vibrations coming from the middle ear via the oval window. As the fluid moves, the cochlear partition, formed by the basilar membrane and organ of Corti, also moves [Henrique, 2014].

Then, thousands of hair cells sense the motion via their stereocilia, and convert that motion to electrical signals, that are communicated via neurotransmitters to many thousands of nerve cells. These primary auditory neurons transform the signals into electrochemical impulses, known as action potentials, which travel along the auditory nerve to structures in the brainstem for further processing.

The terminal of the nerve fibers beneath the hair cells are of two distinct types. The larger and more numerous endings contain many minute vesicles, or liquid-filled sacs, containing neurotransmitters, which mediate impulse transmission at neural junctions. These endings belong to a special bundle of nerve fibers that arise in the brainstem and constitute an efferent system, or feedback loop, to the cochlea [Henrique, 2014].

The smaller and less numerous endings contain few vesicles or other cell structures. They are the terminations of the afferent fibers of the cochlear nerve, which transmit impulses from the hair cells to the brainstem.

The number of outer hair cells in the cochlea is estimated at 12,000, and the number of inner hair cells at 3,500. Because there are 30,000 fibers in the cochlear nerve, there is certainly some overlap in the innervation of the outer hair cells [Joseph E. Hawkins, 2019].

9.4 The Physiology of Hearing

Hearing is the process by which the ear transforms sound vibrations from the external ambient into nerve impulses that are transmitted to the brain, where they are processed and interpreted as sounds. Sounds on the other hand, are produced when vibrating objects, such as the plucked string of an acoustic guitar, induce pressure pulses to make the air molecules vibrate, also known as sound waves.

The ear can distinguish different subjective aspects of a sound, such as its loudness, which is related to the amplitude, and pitch, that is related to the frequency, by detecting and analyzing different physical characteristics of the sound waves. A rise in loudness can prompt physical reactions, such as, increased adrenaline production and hearth rate. This is linked to the subconscious perception of the elevation in sound volume as a possible sign of danger.

In fact, the following associations between music and mood have been observed [Powell, 2010]:

- An increase in *tempo* (rate), loudness (volume), and pitch (frequency) is exciting. On the other hand, a decrease in those parameters has a calming effect.
- If the music is quietly repetitive, which means low volume and repetitive beat, the mood effect is anticipation, or expectation of something to happen.
- The music composed in major keys sounds more exciting, self-confident, or happier than music composed in a minor scale, which is associated with sadness and complex emotions. The major scale inherits the characteristics of completeness from the pentatonic scale.

However, there is a noticeable difference between pitch, a psychological construct of the brain, and frequency, the physical property of a traveling wave. The brain is usually attuned to the overtones of a particular pitch, and fills it in, for example, even if the fundamental is missing. That is the reason a persistent tone is heard by someone who suffers from hearing loss in a certain frequency range [Levitin, 2006].

A sound composed of energy at 100 Hz, 200 Hz, 300 Hz, 500 Hz, and 600 Hz is perceived as having a pitch of 100 Hz, its fundamental frequency. If the signal at 100 Hz is removed by a low-pas filter, the brain continues to perceive this frequency as the pitch. It is called residue pitch, or virtual pitch [Tan et al., 2018].

The human ear is most sensitive to frequencies from 1,000 to 4,000 Hz, but the audible range of sounds extends from about 20 to 20,000 Hz, for healthy young people. Sound waves of higher frequency are referred to as ultrasonic, and can only be heard by other animals.

Loudness is the perception of the intensity of sound, caused by the pressure exerted by sound waves on the tympanic membrane. It is a psychological construct that relates, in a nonlinear manner, how much energy an instrument creates [Levitin, 2006].

The sound perception, in relation to sound measurement, is subject to some problems, such as, subjectivity, the mind state can only be observed indirectly; nonlinearity, the senses respond nonlinearly to the aural or visual aspects of reality, a natural protection of the human body; and nonorthogonality, because perceptual variables influence each other in a non-intuitive manner [Loy, 2011b].

The greater the sound amplitude or strength, the greater the pressure or intensity, and consequently the loudness. The intensity of sound is measured and reported in decibels (dB), a unit that expresses the relative magnitude of a sound on a logarithmic scale.

It is interesting to note that the human senses, including sight, hearing, smell, taste and touch, have a tendency to give logarithmic responses to the corresponding excitations. This indicates some sort of saturation, probably to protect the nervous systems.

The decibel is a unit for comparing the intensity of a given sound with a standard sound that is just perceptible to the normal human ear, at a frequency in the sensitive range of the ear. On the decibel scale, the range of human hearing extends from 0 dB, which represents and inaudible level, to 130 dB, the level at which sound becomes painful [Joseph E. Hawkins, 2019].

For a sound to be transmitted to the central nervous system, its energy experiences certain transformations. The air vibrations excite the tympanic membrane and moves the ossicles of the middle ear. The vibrations are transmitted to the cochlear fluid. Finally, the fluid vibrations initiate traveling waves along the basilar membrane that stimulate the thousands of hair cells of the organ of Corti [Henrique, 2014].

The hair cells convert the sound vibrations to neural impulses in the fibers of the cochlear nerve, which transmits them to the brainstem, from which they are relayed, after some processing, to the primary auditory area of the cerebral cortex, the specialized area of the brain for hearing. When the neural impulses reach a portion of the brain, the listener becomes aware of the sound.

The cochlea also has a second longitudinal membrane, called Reissner membrane, that separates the cochlear duct from the vestibular duct, and, together with the basilar membrane, creates a compartment in the cochlea filled with endolymph [de la Fuente, 2014].

9.5 Transmission of Sound Waves

The outer ear conducts sound waves from the external ambient to the tympanic membrane. The auricle, acts as an antenna collecting sound waves and, with the concha, the cavity at the entrance to the external auditory canal, helps to channel the sound into the canal. The canal enhances the sound that reaches the tympanic membrane.

The resonance enhancement is adequate for sounds of relatively short wavelength, in the frequency range between 2 kHz and 7 kHz, and this controls the frequencies to which the ear is most sensitive, which are important to distinguish the consonantal sounds.

The sounds that reach the tympanic membrane are partially reflected and absorbed to a limited extent. The absorbed sound cause the membrane to move. The acoustic impedance is the predisposition of the ear to oppose the passage of sound, and its magnitude depends on the mass and elasticity of the membrane, and also on the frictional resistance offered by the ossicular chain.

As the tympanic membrane absorbs sound waves, its central portion vibrates as a rigid cone, bending inward and outward. The larger the intensity of the sound waves, the greater the deflection of the membrane and the louder the sound is perceived.

A sound of high frequency induces a fast vibration on the membrane, and causes a high pitch. The motion of the membrane is transferred to the handle of the malleus, the tip of which is attached at the umbo. At very high frequencies the motion of the membrane is complex, and the transmission to the malleus may be less effective [Joseph E. Hawkins, 2019].

9.5.1 Modeling Sound Waves

The direct sound travels directly from the source to the listener, and contains a pure, uncontaminated, version of the original auditory information. The clarity of the sound depends on the direct path. The reflected sound is the result of the return of surfaces, such as, walls, ceilings, floors, and makes the sound unclear, or obfuscate [Tan et al., 2018].

A useful model to explain the effect of the composition of direct and reflected sound is

$$r(t) = \alpha s(t) - \beta s(t - \tau), \tag{9.8}$$

in which $s(t)$ is the direct sound, $s(t - \tau)$ is the reflected sound, that suffers a delay τ with respect to the direct sound, α is the direct path attenuation coefficient, and β is the reflection coefficient. It is assumed that the reflection changes the sound phase by $180°$.

Fourier transform of Equation 9.8 gives [Alencar and da Rocha Jr., 2005]

$$R(\omega) = \alpha S(\omega) - \beta e^{j\omega\tau} S(\omega), \tag{9.9}$$

in which, $R(\omega)$ and $S(\omega)$ are Fourier transforms of the original and received sounds, and $e^{j\omega\tau}$ represents a rotation of the reflected sound by and angle $\omega\tau$ in relation to the direct sound.

In order to plot the spectrum, it is necessary to obtain the magnitude, or modulus, of Equation 9.9,

$$|R(\omega)| = |\alpha S(\omega) - \beta e^{j\omega\tau} S(\omega)| \tag{9.10}$$
$$= |\alpha - \beta e^{j\omega\tau}||S(\omega)|.$$

The ratio between the received and the transmitted Fourier transform magnitudes is known as the transfer function magnitude,

$$|H(\omega)| = \frac{|R(\omega)|}{|S(\omega)|} = |\alpha - \beta e^{j\omega\tau}| \tag{9.11}$$
$$= |\alpha - \beta [\cos(\omega\tau) + j\sin(\omega\tau)]|.$$

Computing the modulus of the expression, gives

$$|H(\omega)| = \sqrt{[\alpha - \beta\cos(\omega\tau)]^2 + [\sin(\omega\tau)]^2}, \tag{9.12}$$

and finally

$$|H(\omega)| = \sqrt{[\alpha - \beta\cos(\omega\tau)]^2 + [\beta\sin(\omega\tau)]^2} \tag{9.13}$$
$$= \sqrt{\alpha^2 + \beta^2 - 2\alpha\beta\cos(\omega\tau)},$$

and the squared magnitude is

$$|H(\omega)|^2 = \alpha^2 + \beta^2 - 2\alpha\beta\cos(\omega\tau). \tag{9.14}$$

The transfer function squared magnitude, that results from the composition of the direct and reflected sounds is plotted in Figure 9.3.

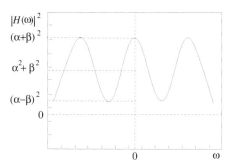

Figure 9.3 The transfer function squared magnitude, resulting from the composition of the direct and reflected sounds.

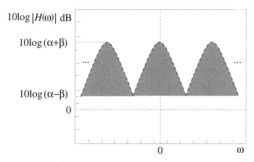

Figure 9.4 The transfer function magnitude resulting from the composition of the direct and reflected sounds, in dB.

The transfer function magnitude resulting from the composition of the direct and reflected sounds, in decibel, is plotted in Figure 9.4. Notice that the spectrum forms a repetitive pattern, indicating that the received sound suffers a frequency-selective fading.

A convenient way to plot the graphic is to use the decibel scale. It is only necessary to take the logarithm, in base 10, of both sides, and multiply the result by 10,

$$
\begin{aligned}
10 \log |H(\omega)| &= 10 \log \sqrt{\alpha^2 + \beta^2 - 2\alpha\beta \cos(\omega\tau)} \\
&= 5 \log \left[\alpha^2 + \beta^2 - 2\alpha\beta \cos(\omega\tau) \right].
\end{aligned} \quad (9.15)
$$

This means that certain frequencies are more heavily affected, that is, undergo a deep fading, by the effect of the sound waves superposition. The specific frequencies depend on the delay τ between the direct and reflected sounds, introduced by the reflection on the surface. The time at which the first reflection is heard, after the direct sound, is called initial time delay gap (ITCG).

The best concert halls present an ITCG of less than 25 ms. This permits the listeners to identify clear beginnings of musical tones, to definitely hear consecutive tones, and to localize the origin of the sound. An ITCG larger than 35 ms makes the hall sound like an arena, and without a sense of closeness [Tan et al., 2018].

Reverberation indicates the perception of how distant is the sound source from the listener, and depends on the dimension of the room or music hall. The reverberation is caused by a multitude of reflected sounds. It has a role in creating a pleasing sound and communicating emotion, and distinguishes the sound produced in a small room from the same music from a large concert hall [Levitin, 2006].

On the other hand, a large number of reflected sounds, or paths, can affect the clarity of the sound. The clarity index (CI) is a measure of the extent to which musical tones sound clear and distinct. It is the ratio of the early sound energy, from the first reflections (arriving within 80 ms of the direct sound) and the late sound energy, the reflections that arrive typically after 80 ms [Tan et al., 2018].

9.6 Synesthesia and Music

The famous circular arrangement of spectral colors appeared, in 1704, in Isaac Newton's (1642–1726) work on *Opticks*. Newton, who was an English mathematician, physicist, astronomer, theologian, author, and philosopher, thought that the spectrum had seven discrete colors, corresponding in some way to the notes of the diatonic scale. The correspondence actually exists, in the sense that one sensory experience may induce another in some individuals, and this is condition is called synesthesia [Sacks, 2008].

Newton's wheel, shown in Figure 9.5, has the colors arranged as arcs, in a clockwise manner, in the order they appear in the rainbow. To each arc limit is assigned a letter. These letters correspond to the notes of the Dorian musical scale, that starts on Re (D), with no sharps or flats.

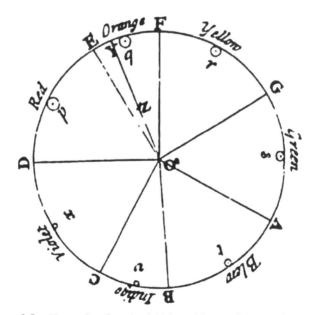

Figure 9.5   Newton's color wheel (Adapted from Wikimedia Commons).

Newton devised this color-music analogy because he thought that the color violet was a sort of recurrence of the color red, in the same way that musical notes recur octaves apart. He introduced the colors orange and indigo at the points in the scale where semitones occur, that is, between Mi (E) and Fa (F), golden, and Si (B) and Do (C), indigo, to complete the octave.

The total list of colors include, in clockwise order, *rubeus* (red), *aureus* (golden), *flavus* (yellow), *viridis* (green), *caeruleum* (blue), *indicus* (indigo), and *violaceus* (violet).

10

Voice Models

Raissa Bezerra Rocha

Federal University of Sergipe, Brazil

> *"Music is the movement of sound to reach the soul for the education of its virtue."*
>
> *Plato*

10.1 The Human Sound Apparatus

The first musical instrument was probably the vocal device. Knowledge of the functioning of the vocal device is fundamental to understand the nature of speech, as it can be viewed as a musical instrument.

The human sound apparatus is a complete and highly complex musical instrument that can translate into ideas the message offered by each sound, through the production of speech. By its very nature, the phoner does not offer a point for the tuning of sounds. Thus, for a complete understanding of how it is possible to use this instrument as efficiently as possible it is also necessary to have a trained ear, that is, sensitive to frequency variations.

It can be said that the vocal device is then a singing musical instrument. However, in order to use the phonation device correctly, as a musical instrument, it is necessary to consider some other aspects such as breathing mode, vocal tract resonance; speech sounds articulation, phonation, vocal projection, body posture, and breaks.

This chapter aims to describe the theoretical basis needed to understand the process of voice production, as well as to relate it to the process of music generation in any musical instrument. So at first this chapter introduces the physiology of the human voice and the whole mechanism of voice production. Emphasizing vocal cord modeling as well as voice synthesis techniques. Finally, it makes a comparison between the spoken and sung voice and describes the vocal device as a musical instrument.

10.2 The Physiology of Human Voice

From the physiological point of view, the voice is produced through three main subsystems, which form the sound apparatus: respiratory, laryngeal, and articulatory [Crovato, 2004; Kent and Read, 1992].

The respiratory subsystem consists of the lungs, trachea, diaphragm, and bronchi, as shown in Figure 10.1. Basically, this subsystem produces an airflow that provides aerodynamic energy to the larynx and articulatory subsystems for the generation of sounds.

The larynx, illustrated in Figure 10.2 is a tubular organ located in the neck above the trachea and below the pharynx. It has three basic functions, protection, breathing, and phonation. Initially, the larynx acts as a protector, preventing foreign elements from reaching the lung with the exception of airflow.

In breathing, more precisely in the exhalation phase, the vocal cords, located in the larynx, are abducted by a set of organs until they touch each other, helping to regulate the gas exchange with the lung and maintaining acid-base balance. Phonation happens when there is vibration of the vocal cords from their tension and longitude changes, besides the enlargement of the glottic opening and the intensity of the breathing effort [do N. C. Costa, 2008; Zitta, 2005].

The articulatory system, as illustrated in Figure 10.3, consists of the pharynx, tongue, nose, teeth, and lips, that is, the vocal tract and nasal tract.

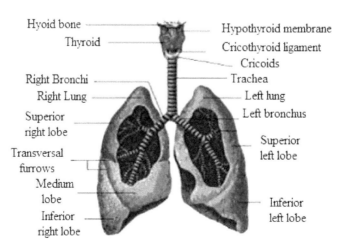

Figure 10.1   Schematic representation of the respiratory subsystem. Adapted from [Crovato, 2004].

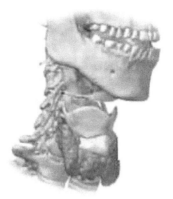

Figure 10.2 Schematic diagram of laryngeal localization. Adapted from [M. E. Dajer, 2006].

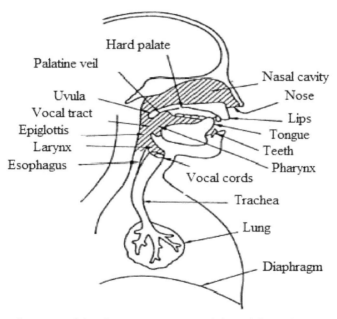

Figure 10.3 Anatomy of the phonatory apparatus. Adapted from [Costa-Neto, 2004; Pacheco, 2001].

10.3 The Speech Production Mechanism

The speech production process consists of three fundamental parts: source of excitement, vocal tract and radiation.

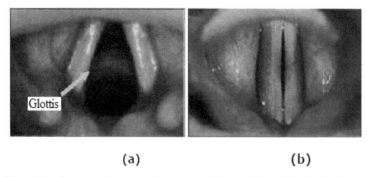

(a)   (b)

Figure 10.4   Vocal cords: (a) open, (b) closed. Adapted from [do N. C. Costa, 2008].

According to [Deller et al., 1993], the voice consists of a wave of acoustic pressure formed from voluntary movements of the human vocal organs.

Its generation begins from the respiratory subsystem, where a flow of air from the lungs is expelled beyond the vocal cords. Basically, sounds are generated by two types of excitement. The first is obtained when there is vibration of the vocal cords in the airflow passage. In the second type of excitation, there is no vocal cord vibration, consisting of a turbulence provided by the passage of air through a constriction in some region of the vocal tract [Selmini, 2008].

As illustrated in Figure 10.4, the vocal cords consist of two pairs of lips symmetrically formed by a muscle and elastic tissue, where the opening between the lips is called the glottis.

The average vibratory pattern of the vocal cords is called the fundamental frequency (F_o), which corresponds to the number of vocal cord vibrations per second. The fundamental frequency is determined based on the length, tension, and mass of the vocal cords, and is given by [Fernandes, 2004].

$$F_o = \frac{1}{L_m}\sqrt{\frac{\sigma_c}{\rho}}, \tag{10.1}$$

in which L_m represents the length of the vibrating membrane of the vocal cord, σ_c the longitudinal tension and ρ the volumetric mass of vocal cord tissue.

For fundamental frequency, it is important to consider the following aspects. The fundamental frequency, or first harmonic, increases with subglottal pressure and vocal fold tissue tension. On the other hand, its value decreases with increasing vocal cord tissue mass. In addition, its value is also

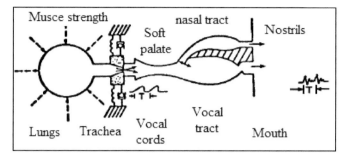

Figure 10.5 Vocal tract model. Adapted from [Rabiner and Schafer, 1978].

related to the vibrant portion of the vocal cord, obtaining a higher value for a short vibration area.

In addition to the vibratory physiognomy, the vocal cords have vertical and horizontal movements, which are represented by an amplitude defined as an extension of the horizontal excursion of the vocal folds during vibration. Range of motion is related to the vibrating portion, stiffness, and mass of the vocal cords, as well as sub-glottal pressure.

The shorter the vibrating portion, the greater the stiffness and mass of the vocal fold, the smaller the range of motion. In contrast, the higher the sub-glottal pressure, the greater the range of vocal cord motion [Zitta, 2005].

The vocal tract, illustrated in Figure 10.5, consists of a resonant filter that begins at the opening between the vocal cords and ends at the lips. The nasal tract begins in the uvula and ends in the nostrils. The vocal tract and nasal tract act as a filter, where the sounds that propagate through them have their frequency spectrum modeled by the filter's frequency selectivity.

Resonant frequencies of the vocal tract are called formants. These frequencies are mainly determined by the shape and size of the vocal tract.

The radiation of the voice is the last step for its generation. The voice is irradiated by both the oral and nasal cavity through the coupling of the tracts from the lowering of the uvula.

10.3.1 Speech Sound Classification

Speech sounds can be divided into two groups: voiced and unvoiced In the voiced generation, the glottis, initially closed, is forced by the pressure of air flow from the lungs, causing its opening.

Thus, during air passage, the vocal cords vibrate in a cyclestationary manner producing a sequence of pulses whose frequency is controlled by the air pressure, tension, and length of the vocal cords. This type of excitement is the basis for the generation of voiced sounds, which in Brazilian Portuguese (BP) are represented by vowels.

Unlike voiced sounds, unvoiced sounds are generated from an excitement formed by the constriction of airflow in the passage of the vocal tract, generating turbulence or wide spectrum noise. For BP, unvoiced sounds are represented by consonants and classified based on four criteria: articulation mode (plosives, fricative, and liquid (rotic and lateral)), as to the point of articulation (bilabial, labiodental, alveolar, palatal, and velar), regarding the role of the vocal cords (voiced and unvoiced) and regarding the role of the oral and nasal cavities (oral and nasal consonants).

Fricative consonants can be divided into two groups according to the type of excitation used in their generation: voiced and unvoiced. Voiced fricatives are characterized by mixed arousal, consisting of vibration of the vocal cords along with a constriction point somewhere in the vocal tract.

On the other hand, in the generation of unvoiced fricatives there is no vibration of the vocal cords. In BP, the voiced fricatives are represented by the phonemes [v], [z], and [j] and the unvoiced phonemes by [f], [s], and [x] [Paranaguá, 2012].

Plosive sounds are due to a complete closure of the vocal tract. They are so named because of their mode of generation, in which there is an explosion corresponding to the sudden release of air after the increased airflow pressure on the vocal cords.

These sounds can also be classified as unvoiced or audible. The unvoiced plosives, characterized by the phonemes [p], [t] and [k] of BP, there is no vocal cord vibration, while in the creation of voiced plosives, such as the phonemes [b], [d], and [g] of PB, there is a small amount of low frequency energy radiated through the throat walls during the total vocal tract constriction period.

The affricated consonants are formed from the combination of fricative and plosive excitation. Like plosives, affricated consonants have their arousal generated by a total constriction at some point in the vocal tract, and upon air release, there is a sound characterized by a noise of friction. In BP, there are two affricated consonants, [T] (unvoiced) and [D] (voiced), which occur by grouping the plosive consonants [t] or [d] followed by the posterior vowel [i].

Nasal sounds are formed through glottal excitation as well as a constriction at some point in the vocal tract. The air is radiated through the nasal tract, but the vocal tract remains attached to the pharynx and the mouth serves as a resonant cavity that emits acoustic energy at a certain frequency. In BP, the three nasal consonants are: [m], [n], and [N].

The lateral consonants, represented in BP by the phonemes [l] and [L], are so named because, after generating vocal cord vibration, the air coming from the lungs travels along the sides of the constriction generated by the tongue.

Unlike the other consonants, the rotic are not generated by a constriction in the vocal tract. Its generation occurs by vibrations in the region of palatal narrowing, as well as the vocal cords. In PB, the rotics consonants are represented by the phonemes [r] and [R].

10.4 Vocal Cord Modeling

The most important mechanisms related to voice production are the vocal cords. Airflow exerts pressure on the glottis causing it to vibrate at a frequency determined by vocal cord mass, length, and tension. The vibratory movement causes air pulses that are modified by the vocal and nasal tract to be subsequently irradiated by the oral and nasal cavities.

The modeling of vocal cord dynamics has been studied for many years and aims to better understand the basic physics of voice and provide diagnosis and treatment for people with voice disorders.

In the literature, it is possible to find some mechanical models that modulate the air passage, describing this mechanism through differential equations.

The first, proposed by Flanagan and Landgraf [Flanagan and Landgraf, 1968] in 1968, describes the movement of the vocal cords by mass-spring-damping mechanical models, as illustrated in Figure 10.6, of mass M, and K and B as the elasticity and stiffness constants.

In order to improve the modeling of vocal cord movement, in 1972 Ishizaka and Flanagan [Ishizaka and Flanagan, 1972] proposed a model in which each vocal cord is represented by two masses, as shown in Figure 10.7, with the vocal cords attached to the laryngeal walls by two nonlinear springs $S1$ and $S2$, and linked together by a linear spring k_C. The model considers that the vocal cords have symmetrical movement in the transverse direction.

In 1994, Titze [Titze, 1994] considered adding a third mass to model vocal cord movement, presented in Figure 10.8. For the masses, the movement perpendicular to the vocal tract is considered. The springs S_1, S_2, and S_3 have nonlinear characteristics and represent vocal cord tensions [E. Cataldo, R. Sampaio and L. Nicolato, 2004].

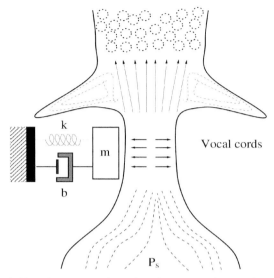

Figure 10.6 Model by Flanagan and Landgraf. Adapted from [Titze, 1994; Brandão, 2011].

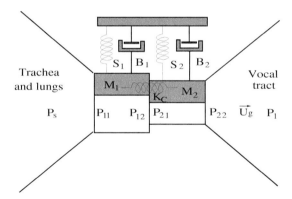

Figure 10.7 Model by Ishizaka and Flanagan. Adapted from [Brandão, 2011].

Other more complex modeling of the vocal cords, aiming to reproduce the irregular vibratory movements, are found in the literature. A finite element model based on the laws of mechanics is proposed to obtain vocal cord oscillation characteristics [Alipour et al., 2000]. The model takes into account a more realistic vocal cord structure, with asymmetry from the point of view of geometry and tension at the vocal cord nozzles, as well as accommodating homogeneities and the anisotropic characteristic of the vocal cords.

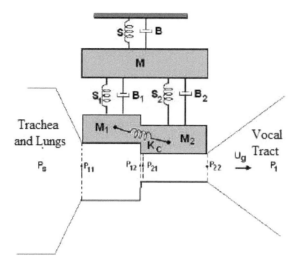

Figure 10.8 Three-mass model proposed by Titze; Adapted from [Brandão, 2011].

In addition, the model allows simulating voice disorders due to vocal cord paralysis.

Vibration irregularity patterns are also taken into account when modeling vocal cords proposed by [Berry et al., 1994; Berry and Titze, 1996; Lucero, 2005]. The dominant spatial modes in the vocal cord finite element model are determined by empirical functions [Berry et al., 1994; Berry and Titze, 1996]. It is observed that complex vibration patterns can be explained by few vocal cord vibration modes, as well as higher-order modes can be estimated by the model in case of irregular phonation.

In [Hüttner et al., 2011], the use of the time-dependent mechanical multi-mass model to estimate pseudo glottal vibrations is reported to generate tissue vibrations in the pharyngoesophageal segment for the production of a sound source in patients whose larynx has been completely removed due to laryngeal cancer. The results indicate that the proposed model achieves better performance compared to similar studies found in the literature, with 97.8 % to 99.7 % correlation rate between the model and the pharyngoesophageal segment.

A synthetic model based on the epithelial layer and lamina propria present in the vocal cords is presented in [Thomson and Murray, 2011]. The results are compatible with those obtained with a finite element model, with the differential of presenting a lower initial glottic tension, being closer to the real human phonation.

The work developed in [Rosa, 2002] presents a mathematical model that represents the larynx during phonation. The objective is to capture physiological phenomena that occur in the larynx during phonation. For this, the larynx muscle tissues are discretized using the finite element method as well as the Navier-Stokes equations. As a result, there is the glottal pulse acquired from different laryngeal geometries with different viscoelastic properties.

In [Patel and Patil, 2014], a study about the length of the vocal cords is performed, based on the Fujisaki model, and relates this characteristic to the tension and fundamental frequency of the vocal cords. The results indicate that the proposed method is not applicable to all speakers, but provide results consistent with the literature [Titze, 1994; Aronson and Bless, 2009], assigning the vocal cord length between 15.2–20.1 mm for men, and 14.3–20.6 mm for women. Furthermore, the authors emphasize that the method can be used to estimate the length of the vocal cords in children, unlike clinical methods, helping in the diagnosis of diseases.

In the literature, there are also studies on pathology detection through vocal cord modeling. In [Marinus et al., 2013], for example, a noninvasive approach to detecting laryngeal diseases is described, as well as discrimination of voice without pathology and voice affected by edema and other diseases such as nodules, cysts and paralysis. For this, cepstral coefficients, neural networks, and Gaussian mixture models are used.

Studies are successful, with a 93% hit ratio in the healthy voice rating and a 94% rate for pathological voice. In addition, the method provided a 76% rate of discrimination for healthy voice and edema voice and 85% for other voice pathologies.

The detection of voice pathologies through images is presented in [Zorrilla et al., 2010]. For this, images are initially obtained by laryngeal video stroboscopy and a contour design *ad-hoc* algorithm is used to obtain robust and rapid image segmentation, and it is possible to identify pathologies and objective measures, such as, among others, the size of the cyst.

The diagnosis of voice pathology is also the subject of study in [Mendez et al., 2007]. In this work, such analysis is proposed by measuring glottis closure and opening angles for various pathologies. The study reports that, in the absence of pathologies, the opening angle is between 35 and 37 degrees, while for paralyzed vocal cords, the angle does not exceed 30 degrees.

In [Zhang et al., 2011], the role of vocal cord stretching in the dynamics of glottic movements during phonation is investigated. It is proposed to include this feature in the two-mass model and it is observed that excessive vocal cord stretching inhibits vibration, but there may be optimal stretching that maximizes vocal cord vibration.

Modeling of the vocal cords using the Port-Hamiltonian systems (PHS) technique is proposed in [Encina et al., 2015]. The authors have succeeded in expressing the *body-cover* model as a PHS and believe that modeling can be employed in acoustic models for voice generation [Story and Titze, 1995].

The effects of polyp mass, stiffness, position and aspects on the natural frequencies and modes of vocal cord vibration are studied in [Greiss et al., 2016] using a finite element code. It is observed that polyp mass is a determining factor in natural frequency and its position has a greater influence on frequency than stiffness.

A voice synthesis model that considers the dynamics of vocal cord movement is that used by linear prediction coding (LPC). The coding method is widespread in the literature and has the characteristic of synthesizing the voice based on parameters extracted from the voice signal waveform [Meireles, 2015].

The LPC is based on linear prediction modeling, which is based on the approximation of a speech signal sample from a linear combination of the previous samples. This principle is related to the voice production model, illustrated in Figure 10.9, which consists of a time-varying linear system with two types of excitation, the periodic, used in the generation of voiced phonemes, and the noisy, for the generation of unvoiced phonemes [Costa, 1994; Teixeira, 1995].

Periodically, for short time intervals in which speech is considered stationary, the predictor coefficients used in the linear combination are computed by minimizing the sum of square differences between the current and the

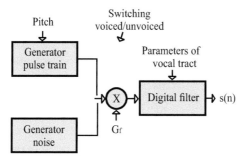

Figure 10.9 Block diagram of the voice production model. Adapted from [Fechine, 1994; Rabiner and Schafer, 1978].

linearly predicted speech sample. These coefficients can be obtained by LPC analysis, called LPC coefficients, or by techniques derived from this analysis. The coefficients used include: LPC, cepstral, weighted cepstral, cepternal delta, and weighted cepstral delta coefficients.

For speech generation, LPC has two types of excitation. To generate the voiced sounds, the LPC assumes that the vocal cords vibrate periodically. Thus, it makes use of a uniformly spaced pulse train as excitation. In contrast, unvoiced sounds are generated from noisy excitement. In addition to the excitation type, the LPC performs voice synthesis with period of voice excitation, gain factor and vocal tract parameters that are used as predictor coefficients [Sotero, 2009; Silva, 2012; Rabiner and Juang, 1996].

In speech signal synthesis, the LPC encoder uses periodically received parameters, which contain information about the model and the nature of the excitation, to reconstruct speech through a mathematical model given by Equation 10.2, constructed from predictor gain and coefficients, in which each sample is generated from the current system input, plus a linear combination of the predicted vocal tract output [Furui, 1985].

$$y[n] = \sum_{h=1}^{p} a_h y[n - h] + G_f x[n], \qquad (10.2)$$

in which $y[n]$ is the n-th output sample, a_h is the h-th predictor coefficient, G_f is the gain factor, $x[n]$ is the input sampled at a time n, and p is the model order.

Linear prediction coding is the basis for several relevant coding algorithms, such as regular pulse excited-long term predictor (RPE-LTP), code excited linear prediction (CELP), vector sum excited linear predictive (VSELP), algebraic code excited linear predictive (ACELP), Qualcomm Code excited predictive (QCELP), adaptive multi-rate narrowband (AMR-NB) and adaptive multirate wideband (AMR-WB) [Silva, 1996; G.729, 1996; Bessette et al., 2002; Salami et al., 2006; Maia, 2000].

10.5 Speech Synthesis Techniques

Speech synthesizers are systems capable of artificially reproducing voice. The synthetic voice can be generated by the concatenation of acoustic units or by systems that incorporate the modeling of the vocal device, as a mechanism of vibration of the vocal cords and characteristic of the filter represented by the vocal tract.

Speech synthesis is widely used in text-to-speech (TTS) systems, with a variety of applications, such as phone text query, call center electronic attendance, accessibility systems, and voice coders, among others.

This section presents the main syntheses disseminated in the literature, which are articulatory, concatenation, and formant.

10.5.1 Articulatory Synthesis

Voice production through articulatory synthesis occurs through the modeling of the human phonatory apparatus. In this case, the characteristics of the articulators that participate in voice production, as well as the glottal opening movement, such as vocal cord tension and lung pressure, are taken into account.

The modeling of the articulatory synthesis requires knowledge of a set of parameters, such as the lip-opening area, the tongue blade constriction, the opening to the nasal cavities, the average glottal area, and the rate of expansion or contraction of the volume in the region of the vocal tract corresponding to the pharynx. However, such parameters are usually obtained in 2D from X-ray analysis, but the actual vocal tract is naturally 3D, making it difficult to optimize this type of template [Araújo, 2015; Brito, 2007].

The paper by [Howard and Huckvale, 2005] proposes an articulatory synthesis using the supervised regression technique and data obtained from a MOM-based generator to train a model that links acoustic voice representation with articulatory parameters of the synthesizer. The synthesizer was evaluated by hearing tests, as well as spectrogram analysis, obtaining good performance by synthesizing voice signals used in the model training.

Imitation of the voice using articulatory synthesis with modeling with neural networks is described in [Philippsen et al., 2014]. The model is built to synthesize syllables and is capable of articulatory-acoustic mapping, and vice versa, for consonant-vowel sequences, including co-articulatory effects.

10.5.2 Concatenation Synthesis

Concatenative synthesis is characterized by producing sound by joining segments corresponding to acoustic units. This type of synthesis is performed in three steps, presented in Figure 10.10.

The first stage consists of the formation of the acoustic unit bank obtained by segmenting previously recorded phrases. In a concatenation synthesis, choosing the size of the units to be used in the synthesis process is one of the most important decisions, as it must represent a compromise between

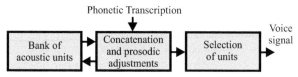

Figure 10.10 Block diagram of speech synthesis by concatenation. Adapted from [Costa-Neto, 2004].

intelligibility and required naturalness. There are several possibilities of sizes and quantities that can be used.

One of the segments that can be used in concatenation synthesis is the diphone, unit formed by a pair of phones. It starts halfway through the first handset and ends halfway through the next handset. Its advantage is that it contains entirely the transitions between the phones. However, diphones include only part of the various co-articulatory effects of the spoken language, which justifies even partial use, of larger units such as triphones.

Triphones are segments that include an entire phone and its transitions left and right. However, due to the large amount of triphones present in the Portuguese and other Latin languages, these units are used as a complement, for special sound cases, to banks based on smaller units [Taylor, 2009; Dutoit, 2011].

Other units that can be used in concatenation synthesis are the halves of phones, syllables, semi syllables, words, and phonemes. Half of the phones extend from the boundary between phones to the midpoint, or extend from the midpoint to the end of the phones. However, this unit when used in isolation presents difficulty of representation of coarticulation.

The syllables can be considered natural units, since they present the coarticulation between the phonemes that form them and are more important than the coarticulations present in the intra-syllable segments. However, the absence of these coarticulations decreases the quality of the synthesized signal. Another disadvantage of these segments is their large amount in the Portuguese language, for example, which makes it difficult to build a bank using this type of segment [da Silva, 2011].

Based on the same phonological principles as syllables, semi-syllables are formed by dividing the syllables into two partially overlapping parts, with the syllable peak (nucleus) belonging to both parts. An disadvantage of using this type of segment is that it is not always possible to neglect the interaction that occurs between segments belonging to different syllables [Simões, 1999; Latsch, 2005].

In addition to the exposed concatenation units, the synthesis of a speech signal can also be performed with words or phrases. The disadvantage of this type of unit is the large number required in an unrestricted synthesis system, that is, that synthesis that is not restricted to a set of words or phrases.

Concatenation synthesis can also be performed using phonetic segments. Its advantage is the small number of units present, for instance, in the Portuguese language, which allows the use of a small voice bank. However, the synthesis using these units has an unstable behavior, which oscillates between synthesized speech with a high degree of naturalness and speech synthesized with unpleasant distortions [Morais, 2006].

This is due to the fact that the coarticulation points can be performed at the boundaries of the phones, making it difficult to accurately represent the coarticulation effect, which requires multiple samples of the same unit in different contexts (allophones) [Latsch, 2005].

In a concatenation synthesis, it is necessary to take into account the variation to which the concatenation units are subjected according to the position occupied within a sentence or to the applied intonation. Thus, to maintain a correct and natural intonation, it would be necessary to consider all variants of the phoneme as concatenation units, which would be chosen according to grammatical or semantic rules.

After choosing the size and types of units to be used in the synthesis, the next step is to select them from the phonetic transcription of the phrase to be synthesized. Finally, the concatenation of the acoustic units is performed, with the possibility of adjusting the energy, duration, and fundamental frequency.

However, some disadvantages may occur in concatenative synthesis: spectral envelope discontinuities and amplitude, pitch, and phase discontinuities between the segments. Spectral discontinuities occur when the formants of adjacent segments do not have the same values and are mainly related to coarticulation. This problem can be mitigated by smoothing the edges of the segments [Costa-Neto, 2004; Dutoit, 2011; Klabbers, 2000].

10.5.3 Formant Synthesis

Formant synthesis is based on the source-filter theory. Its accomplishment occurs through three components: excitation source, vocal tract filtering characteristics, and radiation characteristic to the external environment.

The excitation source for sound signals consists of a pulse generator evenly spaced for a time equal to the pitch period. In the production of unvoiced sounds, the excitation is represented by a noise generator.

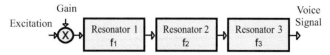

Figure 10.11 Block diagram of a formant synthesizer with series or cascade configuration. Adapted from [Costa-Neto, 2004].

The vocal tract is a filter whose transfer function is composed of resonators whose purpose is to model the frequency and bandwidth of each formant. The transfer function of a resonator is given by

$$R_n(z) = \frac{a_{1n}}{1 - a_{2n}z^{-1} - a_{3n}z^{-2}}, \tag{10.3}$$

in which a_{1n}, a_{2n} and a_{3n} are coefficients related to central resonant frequency, f_n, and formant bandwidth B_n, given by

$$a_{3n} = -e^{2\pi B_n T_s}, \tag{10.4}$$

in which T_s is the sampling period in seconds,

$$a_{2n} = 2e^{-2\pi B_n T_s} \cos(2\pi f_n T_s), \tag{10.5}$$

and

$$a_{1n} = 1 - a_{3n} - a_{2n}. \tag{10.6}$$

A formant synthesizer can be built in series or parallel. Figure 10.11 illustrates the series association of resonators, which has the advantage that it does not require a specific gain for each resonator, unlike parallel association, as illustrated in Figure 10.12. Serial association has the disadvantage that the transfer function is not adequately modeled for the production of fricative and plosive sounds [Pacheco, 2001].

In parallel association, the excitation signal is applied to all resonators and their outputs are summed. In this case there is an individual control of the gain and bandwidth of each formant. This setting produces nasal, fricative, and plosive sounds with better quality than the cascade structure. In contrast, its transfer function is not adequately modeled for the production of vowels.

To synthesize speech, formant synthesizers periodically receive information such as amplitude, fundamental sound signal frequency, and formant frequencies and bandwidths. An example of a widespread formant synthesizer in the literature is Klatt [Klatt, 1980; Klatt and Klatt, 1990a], controlled by 39 parameters that are updated every 5 ms. In recent years, Klatt's synthesizer has been used in [Anumanchipalli et al., 2010] in the generation of synthetic voice.

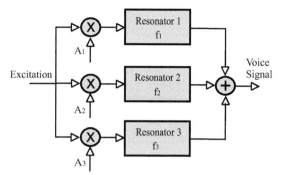

Figure 10.12 Block diagram of a formant synthesizer with parallel configuration. Adapted from [Costa-Neto, 2004].

10.6 The Spoken and Sung Voice

The singing voice and the spoken voice have their own characteristics that differentiate them from each other.

The singing voice is produced from the same mechanism of production of the spoken voice, that is, source of excitation, vocal cords, vocal tract and radiation. However, in order to give rise to the singing voice, some adjustments to the driving mechanism need to be made. According to [Behlau and Rehder, 1997] the spoken voice and the singing voice differ in terms of breathing mode, resonance, articulation of speech sounds, phonation, vocal projection, body posture and pauses.

Breathing is the starting point for the production of spoken and sung voice. In the spoken voice the breath is naturally commanded by the individual according to the intonation and emotion that wishes to produce his or her speech. In this case, for speech production, the lungs are expanded faster compared to the breath generated without the purpose of voice production.

In sung voice, breathing is a process programmed according to the musical portion you wish to sing. Inspiration is performed quickly and is predominantly oral. The lungs are expanded using all the chest walls and with a larger air volume compared to the lung air volume used for spoken voice. Exhalation is also performed in a controlled manner to keep the ribcage expanded longer until virtually all air is expelled.

Phonation is the second stage in the production of both voices and refers to the physiological behavior of vocal cords in the production of spoken and sung voices. According to [Rocha, 2017] in the production of spoken voice, the vocal cords are moved from an electrical signal from the brain, which provides them with a cyclestationary movement when excited by the airflow from the lungs.

The glottal pulse generated at each vocal cord opening and closing can be represented from some models found in the literature such as Rosenberg [Degottex, 2010], Fant [Fant, 1979], Liljencrants-Fant (LF) [Fant et al., 1985], and Klatt's [Klatt and Klatt, 1990b].

In spite of the difference in the number of parameters between the models, they have similarities in their characteristics, such as, representing the always positive or null glottal pulse, considering the almost periodic glottal flow and as a continuous function in time. These pulses have a longer opening period than the closing period. Figure 10.13 illustrates the glottal pulse behavior of the Liljencrants-Fant model, and Figure 10.14 illustrates the vocal cord behavior at the moment of phonation.

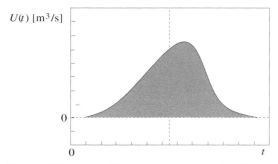

Figure 10.13   Liljencrants-Fant glottal pulse model [Gobl, 2003].

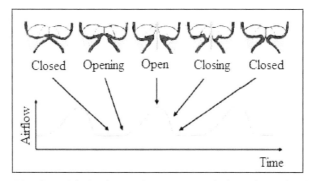

Figure 10.14   Vocal cord behavior at the moment of phonation [dos Santos, 2009; Dias, 2012].

In the production of spoken voice, the larynx presents moderate movement, usually associated with variations of emotion and intonation of the generated phrases.

Unlike the pulse generated by vocal cords in spoken voice production, the pulse generated in singing voice production has the characteristic of having a longer closing period than opening. This feature is important as it allows you to generate longer and more acoustically defined sounds. In this case, the larynx, where the vocal cords are located, usually remains in the low position for both high and low frequencies.

The next step in voice production is the vocal tract, responsible for resonance. The vocal tract is a filter that alters the glottal flow from the vocal cords according to resonant or formant frequencies. In spoken voice, the resonance is medium, while in the singing voice, it is high and concentrated in the upper part of the vocal tract.

Another feature that differentiates spoken and sung voice is the vocal quality. The quality of the spoken voice is related to the nature of the speech and the characteristics of the speaker, such as tone, emotions, among others. The singing voice requires training and specific prior adaptations according to what one wishes to produce, making the quality of the singing voice more stable than the quality of the spoken voice.

Thus, the vocal quality of the singing voice is not directly linked to the individual's personal characteristics, but the characteristics of what one wishes to sing, such as the repertoire and musical style.

In sung voice the vowels are longer than consonants, which further improves their quality. Thus, the articulation of some sounds can be altered according to the tonal aspects of the music and the musical passage. In spoken voice, the duration of vowels and consonants are related to what one wishes to speak, and the physical and regional characteristics of the individual. In this case, the junctions must be accurate so that the identity of the sound does not change.

Other important features are pause, speed, and pace. In the spoken voice, these three factors are related to the characteristics of the speaker, that is, their regionality, language, personality, and even with the intensity desired in the emotion of another individual. In sung voice, the three factors are programmed and are closely linked to the music and harmony.

As a result of the different characteristics of the voice production process, the singing voice also presented some differences in its frequency spectrum. Figures 10.15 and 10.16 illustrate, respectively, the power spectral densities

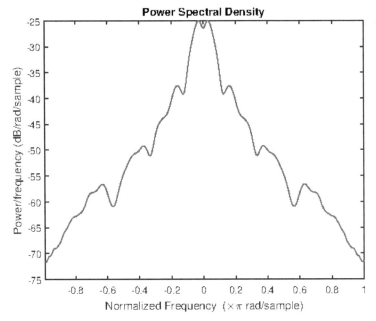

Figure 10.15   Power Spectral Densities of a spoken voice.

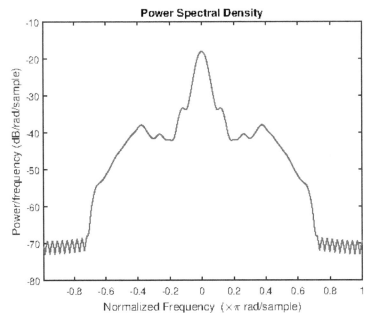

Figure 10.16   Power Spectral Densities of a singing voice.

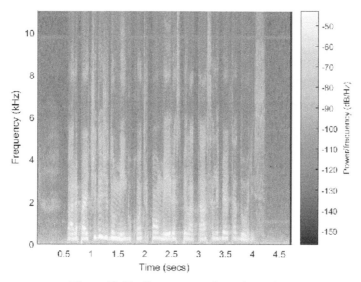

Figure 10.17 Spectrogram of a spoken voice.

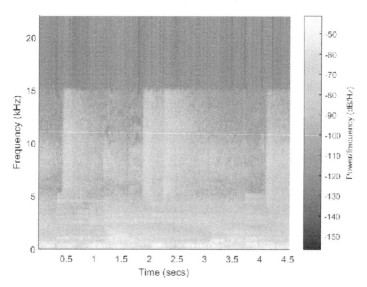

Figure 10.18 Spectrogram of a singing voice.

of a spoken voice and a singing voice and Figures 10.17 and 10.18 illustrate, respectively, the spectrograms of a spoken voice and a singing voice.

From a comparison between the spectrograms illustrated in the figures, it can be seen that the singing voice has a higher energy in the mid-frequency regions of the spectrum, due to differences in laryngeal positions in the sung voice generation [de la Fuente, 2014].

10.7 The Voice Apparatus as a Musical Instrument

The vocal device has the ability to generate sounds as varied and beautiful as a musical orchestra. As mentioned, speech is produced by the phonatory device, which is formed of three independent stages: the source of excitation, the vocal tract (resonance), and the radiation region. Therefore, the vocal device can be visualized as a musical instrument whose music is produced by speech.

To make a sound, each musical instrument also needs three basic structures: a source, a resonator, and a radiator. The source generates a specific frequency, while the resonator acts as an amplifier, reinforcing the produced sound. The beam is represented by a passage that transmits the sound out of the musical instrument.

The structure of the vocal device can be compared to the structure of any musical instrument. For example, the violin. In a violin, the strings represent the sound source.

In the phonatory apparatus, the source consists of air that comes from the lungs and vibrates the vocal cords according to the pronounced phoneme. In the violin, the resonator is represented by the cavity of your body, applying the generated sound. In speech production, the resonator is the vocal tract that modifies the glottal flow according to formant frequencies. Finally, the radiator of the phonatory apparatus is the mouth and nostrils, while the violin radiates the sound through the lateral openings located in its upper part.

A characteristic of the phoner is that it generates a signal that has several frequencies, which is speech. Each frequency is seen as a musical note. In this way, the vocal device can be seen as an instrument without fixed tuning, being used to generate melodies in various tones.

Besides being a complete musical instrument, the vocal device is quite complex. Thus, its precise tuning can only be adjusted by the human ear, which is trained to be sensitive to the frequency variation that characterize the musical notes.

11

Scale Formation

"Music is the mediator between the spiritual and the sensual life."
Ludwig van Beethoven

11.1 The Emergence of the Scales

The human ear is sensitive to acoustic waves in a wide range of frequencies, and can detect very small frequency variations, a resolution of less than 0.5% in the discernible threshold. On the other hand, the western music is based on scales that have frequency transitions of 20 times that resolution.

To find a reason for the emergence of the scales, or quantized systems of tones, is a complex task. First, the brain needs some time to process the musical note, which makes difficult the use of a continuous system. Second, most musical cultures have always used different scales, which are associated to that particular tradition. Third, most primitive musical instruments had fixed pitches. Finally, the existence of scale is also associated to the polyphonic music.

In neuropsychological terms, the existence os scales can be justified by the facility that it offers the brain to process, identify, and store a melody that is formed by a temporal sequence of discrete pitch values, which are related in some way, as in a harmonic series [Roederer, 1998].

11.2 Definition of Scale

A scale is a family of pitches arranged in an ascending or descending order. Each scale follows a set pattern of intervals. The key of the scale is defined by the first, or root, note of the scale [Buck, 2014].

The *key* is the central note, chord, or scale of a musical composition or movement. The key signature represents a series of sharps or flats written on a musical staff to indicate the key of a composition.

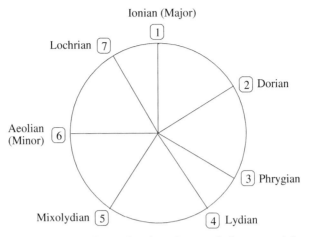

Figure 11.1   The modes produced starting a scale from several degrees.

Table 11.1   The diatonic scale and the interval order

Diatonic degrees	1 2 3 4 5 6 7
Interval orders	2 2 1 2 2 2 1

Table 11.2   The minor scale and the interval order

Minor degrees	1 2 3 4 5 6 7
Interval orders	2 1 2 2 1 2 2

The *tonic* is the first note of a scale or key, while tonic key is the home key of a tonal composition.

A *mode* is a scale or key used in a musical composition. Major and minor are modes, as are ancient modal scales found in western music before c.1680, such as, Ionian (Major), Dorian, Phrygian, Lydian, Mixolydian, Aeolian (Minor), and Lochrian. The modes produced starting a scale from several degrees are shown in Figure 11.1.

For instance, the Major, Ionian, or diatonic, scale has the interval order (number of semitones between the notes of the scale) displayed in Table 11.1. This is represented by the amplitudes of the arcs in the circle.

The diatonic scale, which is a model of sound organization in the Western culture, is composed of seven notes, that contains five tones (Do–Re, Re–Mi, and Fa–Sol, Sol–La and La–Si) and two semitones (Mi–Fa and Si–Do) [Wisnik, 2017].

The minor, or Aeolian, scale interval order is displayed in Table 11.2. It can be obtained from the circle starting from degree 6.

11.3 Main Scales

Tonality is related to music centered around a home key, based on a major or minor scale. The popular music, for example, is typically tonal, centered around a single musical pitch called *tonic* [Wyatt and Schroder, 1998].

11.3.1 Major Scale

A major key is a type of music based on a major scale, which is traditionally considered happy sounding, while a minor key is a kind of music based on a minor scale, which is traditionally considered sad sounding.

The major scale is a family of seven alphabetically-ordered pitches within the distance of an octave, following an intervalic pattern matching the white keys from "C" to the one octave above "C" on a piano. The formula for the major scale is shown in Figure 11.2, in which W means whole tone, and H means half-tone intervals [Wyatt and Schroder, 1998].

Regarding the qualitative nature of the intervals, the major third (M3) has a distance of four half-steps, or four semitones, and offers a jovial, positive, and assuring character to the part. The major sixth (M6) has a distance of nine half-steps, and is a large consonant leap. The major seventh (M7) has a distance of 11 steps, and is a large dissonant leap that creates a dramatic tension [Franceschina, 2015].

The minor intervals are produced when the distance between two tones is shortened by one half-step in a major interval. The minor third (m3) has a distance of three half-steps and is consonant, but somehow darker that M3, because it appears unresolved and evokes some tension, an unfulfilled desire and a feeling of longing. The minor sixth (m6) has a distance of eight half-steps, is a consonant leap, and is used to express emotion and drama. The minor seventh (m7) has 10 half-steps, and is a usual blues interval in music,

Figure 11.2   The formula for the major scale.

Figure 11.3   The formula for the minor scale.

presenting a movement directed to an anticipated resolution [Franceschina, 2015].

11.3.2 Minor Scale

The minor tonality is more complex than the major one, mainly because it involves three scales, instead of one. The basic scale of the minor system is called natural minor, and comes from the ecclesiastic Aeolian mode. This scale starts in the sixth degree of the major scale and share with it the same alterations. Because of this, those tonalities are known as relative [Juanilla, 2014].

The minor scale is family of seven alphabetically-ordered pitches within the distance of an octave, following an intervalic pattern matching the white keys from "A" to an octave above "A" on a piano. The minor scale represents a darker, melancholic sound [Schonbrun, 2014].

The formula for the minor scale is shown in Figure 11.3, in which W means whole tone, and H means half-tone intervals [Wyatt and Schroder, 1998].

11.3.3 Other Scales

A whole-tone scale is made of six whole steps that avoids any sense of tonality. Example: C D E F# G# A#.

A song is a small-scale musical work that is sung. A German song is a called *Lied*, a French song is a *chanson*, and an Italian song is a *canzona*. A song cycle is a set of poetically-unified songs (for one singer accompanied by either piano or orchestra. On the other hand, a through-composed form is a song form with no large-scale musical repetition.

The pentatonic scale is usually defined as a folk or non-Western scale having five different notes within the space of an octave. But the pentatonic scale is versatile and can be used with different music styles. The blues scale is almost identical to the pentatonic minor scale, and the only difference is the insertion of the "blues" note, or a flat fifth ($\flat$5th), to name the scale [Buck, 2014].

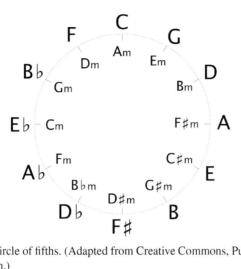

Figure 11.4 The circle of fifths. (Adapted from Creative Commons, Public domain, created by Joseph Harrington.)

11.4 The Circle of Fifths

The circle of fifths is a pattern that is used to study the relationship between the keys. It is a sequence that begins with a tonic, followed by a dominant, and so on, as shown in Figure 11.4.

11.4.1 Chroma and Pitch Perception

The studies of musical expectation, constructed on previous work in music perception, targeted on finding a way to represent pitches to model the human perception of them. The chroma circle represents the 12 pitch classes as a circle of contiguous pitches that are each one a half-step distance from the other.

In order to explain the perception of pitch, Moritz Wilhelm Drobisch (1802–1896), a German mathematician, logician, psychologist, and philosopher, proposed a geometric model, a helical transformation of the chroma circle, shown in Figure 11.5 [Nikolaidis, 2011].

In the transformation, shown in Figure 11.6 with a few particular notes, Drobisch extended the chroma circle's representation of relative distance in pitch class to also represent change in pitch height.

Using this new model, that can be mathematically expressed using Euler's formula, a distance relationship could be shown between any two pitches, not just the 12 pitch classes of the chromatic scale [Henrique, 2014].

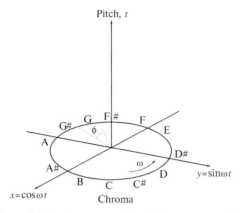

Figure 11.5   A representation of chroma perception.

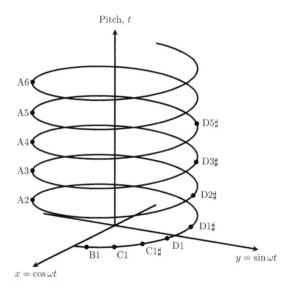

Figure 11.6   A helical representation of pitch perception.

A complex exponential is a useful mathematical model to express the helical representation of pitch,

$$r = x + jy. \tag{11.1}$$

It is possible to write the orthogonal coordinates in polar form, $x = \cos \omega t$ and $y = \sin \omega t$, and the equation becomes

$$r = e^{j\phi} = \cos \phi + j \sin \phi, \tag{11.2}$$

in which, $\phi = \omega t$, is the chromatic angular position or displacement, $\omega = 2\pi f$ is the angular velocity, or the speed of change, and t represents the pitch height.

Therefore, the famous Euler's equation can be used as a mathematical model for some of the musical parameters,

$$e^{j\omega t} = \cos \omega t + j \sin \omega t. \qquad (11.3)$$

With the introduction of the intensity, or loudness, parameter L, which is the radius of the chromatic circle and represents the music dynamics, the model becomes

$$s(t) = Le^{j\omega t} = L\left[\cos \omega t + j \sin \omega t\right], \qquad (11.4)$$

in which $s(t)$ is the equation for a helix, and represents the organized sound, in the model proposed by Drobisch.

It is important to recall that the frequency of the note, perceived as a chromatic change or as the pitch, is a continuous parameter, and that the quantization of the scale is just a useful and convenient convention to write music.

Euler was much occupied with music during his entire life. Nicholas Fuss, his student, son-in-law, and secretary, recorded that "Euler's chief relaxation was music, but even here his mathematical spirit was active. Yielding to the pleasant sensation of consonance, he immersed himself in the search for its cause and during musical performances would calculate the proportion of tones" [Pesic, 2013].

12

Consonance and Dissonance in Music

"Music is the one incorporeal entrance into the higher world of knowledge which comprehends mankind but which mankind cannot comprehend."

Ludwig van Beethoven

12.1 Definition

There is some subjectivity regarding the sound perception by different people, but, for some specific simultaneous tones, the perceived sounds can be classified as consonant or dissonant [Arbonés and Milrud, 2012].

An important concept related to sound perception is the critical bandwidth (CB), a range of frequencies that evoke a similar response in the auditory system. Because frequency is a continuous variable, it is difficult for the auditory system to respond individually to every possible value, therefore, it responds similarly to frequencies that are in a short range, or critical band. An example of this is the subjective effects of musical consonance and dissonance [Tan et al., 2018].

12.1.1 Consonance

The sounds are said to be consonant if they are pleasant, stable, and agreeable. There are certain common types of consonances, that may include [Ball, 2010]:

- Perfect consonances
  - unisons and octaves and
  - perfect fourths and perfect fifths.
- Imperfect consonances
  - major thirds and minor sixths and
  - minor thirds and major sixths.

12.1.2 Dissonance

If the sound has the impression of tension, is unpleasant or unstable, and causes a desire to be resolved to consonant intervals, it is considered dissonant Dissonances can be classified, regarding atonal music, as resolution, in tonal music theory, is the move of a note or chord from dissonance to consonance [de Mattos Priolli, 2013a].

- Sharp dissonances
  - minor second and
  - major seventh.

In tonal music, non-diatonic intervals, the following are commonly dissonant:

- Soft dissonances
  - diminished and
  - augmented.

The perceived dissonance is maximal when there are overtones that fit within the same CB, while not being identical in frequency. The greatest dissonance is associated with one fourth of the bandwidth, which occurs for highly dissonant intervals, such as, the tritone, for example Sol (G) and Do# (C#). The overtones for consonant intervals are identical in frequency, are double in frequency, or do not fit within the critical bandwidth, for instance, Sol (G) and Re (D) [Tan et al., 2018].

12.2 Modeling Consonance and Dissonance

Music organization is, in a certain sense, similar to mathematics; and a musical style is akin to a language development, because physiological, physical, and neurological characteristics can impose limits to the development of a musical style.

12.2.1 Beating and Dissonance

It is well known that two notes played simultaneously, but with slightly different frequencies, produce an effect that is perceived as a beating sound. The musical term for this effect is *tremolo*. When the frequency difference increases past 10 Hz, the tremolo effect disappears and the tone sound becomes rough and unpleasant, or dissonant [Loy, 2011b].

The effect can be observed considering the sum of two sinusoids, with angular frequencies ω_A and ω_B, as follows

$$s(t) = \cos(\omega_A t) + \cos(\omega_B t). \tag{12.1}$$

Application of a trigonometric property leads to

$$s(t) = 2\cos\left[\frac{(\omega_A + \omega_B)t}{2}\right] \cdot \cos\left[\frac{(\omega_A - \omega_B)t}{2}\right]. \tag{12.2}$$

Therefore, the combination of the sinusoids produces a modulated signal, with a carrier frequency that is equivalent to the sum of the original frequencies, $\omega_A + \omega_B$, and a modulating frequency that is the difference of the original ones, $\Delta\omega = \omega_A - \omega_B$.

Dissonance is usually defined by the amount of beating between partials, that are called harmonics or overtones, when occurring in harmonic timbres.

The effect of frequency beating for two sinusoidal signals, that present a small frequency difference, is illustrated in Figure 12.1. The composite signal presents nodes, because of the destructive interference, and peaks, when the interference is constructive.

Figure 12.2 illustrates the effect of frequency beating for two signals with a greater frequency difference. The signal has a noise-like appearance.

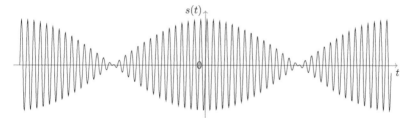

Figure 12.1 Effect of frequency beating for two signals, with a small frequency difference.

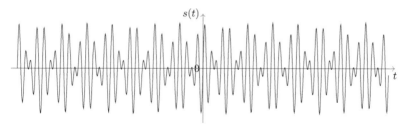

Figure 12.2 Effect of frequency beating for two signals with a frequency difference greater than 10%.

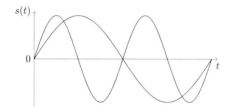

Figure 12.3　Precession time for the Octave, ratio = 2:1. The precession time is two.

If the frequency difference is above 20 Hz, the ear begins to hear two distinct, but irregular, tones. The increase in frequency difference, as it approaches a major third, will eventually lead to the perception of two separate and clear tones.

12.2.2 Precession Time and Consonance

The precession time, the time required for the signals to coincide in an interval, can be used to indicate consonance. The idea was first proposed by Giovanni Battista Benedetti (1530–1590), a mathematician and philosopher who published his first book, *Resolutio*, at Venice, when he was only 22 years old. The book concerns the general solution to all of the problems in Euclid's Elements.

Benedetti related interval consonance to the frequency of waveform coincidence between two tones. He argued that the wavelengths of increasingly consonant music intervals coincide more frequently than those of more dissonant intervals [Loy, 2011b].

In fact, the wavelengths of consonant music intervals coincide more frequently, as shown in the illustrations. The precession time for the octave is illustrated in Figure 12.3, that shows two sinusoids, with a frequency ratio of 2:1. The precession time for this case is two.

Figure 12.4 shows two sinusoids, with a frequency ratio of 3:2, to illustrate the precession time for the perfect fifth interval. The precession time for this case is six.

Figure 12.4 shows two sinusoids, with a frequency ratio of 4:3, to illustrate the precession time for the perfect fourth interval. The precession time for this case is 12.

12.2.3 An Additive Dissonance Metric

Recalling that, in music theory, an interval is the distance between two pitches and that the smallest interval in Western music is a half-step. There are three

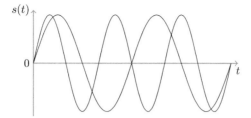

Figure 12.4 Precession time for the fifth, ratio = 3:2. The precession time is six.

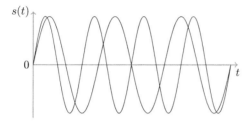

Figure 12.5 Precession time for the fourth, ratio = 4:3, precession time = 12.

basic types of intervals regarding consonance: the perfect, the imperfect, and the dissonant. Imperfect intervals can be either major or minor.

The additive dissonance metric (ADM), the sum of the numerator and denominator of the ratio between tone frequencies, is monotonically related to dissonance. Regarding the perfect intervals, for example, for the unison (1:1) the value of ADM = 2, for the octave (2:1) ADM = 3, for the fifth (3:2) ADM = 5, and for the fourth (4:3) ADM = 7. Perfect intervals have only one basic form, and they form a decreasing order of consonant intervals.

Imperfect intervals have two basic forms, they can either be a major or a minor interval. For the imperfect intervals, in increasing order of dissonance, one finds the major sixth (5:3), that has the value of ADM = 8, the major third (5:4), that has ADM = 9, the minor third (6:5), that has ADM = 11, and the minor sixth (8:5), that has ADM = 13.

The dissonant intervals follow the same criterion. The major second (9:8) has the value of ADM = 17, the major seventh (15:8) has ADM = 23, the minor seventh (16:9) has ADM = 25, the minor second (16:15) has ADM = 31, and the tritone (64:45) has ADM = 109 [Loy, 2011b].

As expected, the interval qualities can be described as major, minor, harmonic, melodic, perfect, augmented, and diminished. When a perfect interval is lowered by a half-step it becomes diminished. When it is raised a half-step it becomes augmented.

If a major imperfect interval is lowered a half-step it becomes a minor interval. If it is raised a half-step it becomes augmented. If a minor interval is lowered by a half-step it becomes diminished. If a minor interval is raised a half-step it becomes a major interval.

12.2.4 Frequency Ratios and Consonance

In addition, pairs of frequencies that reduce to fractions, such as, $\frac{n+1}{n}$, for small values of n, are the ones that usually sound more consonant.

Table 12.1 shows the relation between the ratio of two frequencies in an interval and the perception of the interval's consonance or dissonance, indicated in the last column.

The evolution of harmony has been, in a sense, the redeem of the dissonance. The intervals that used to be considered dissonant were avoided, such as, the augmented fourth interval, for centuries called *Diabolus in Musica*, which means the devil in music, because it sounded particularly unpleasant.

Against the mainstream of music, Leonard Bernstein (1918–1990), an American composer, conductor, author, music lecturer, and pianist, used this interval to built the theme "Maria" in the musical Wet Side Story, with great success [Juanilla, 2014].

12.2.5 Euler and Music

Leonhard Euler devised a mathematical formula to deal with the dissonance question. In 1739, Euler defined an arithmetic function that established a

Table 12.1　Relation between the ratio of two frequencies in an interval and the perception of the interval's consonance or dissonance

Name	Ratio	Approx. fraction	Perception
Unison (C)	1	1/1	Consonant
Minor second (C♯/D♭)	1.059463	16/15	Dissonant
Major second (D)	1.122462	9/8	Dissonant
Minor third (D♯/E♭)	1.189207	6/5	Consonant
Major third (E)	1.259921	5/4	Consonant
Perfect fourth (F)	1.334840	4/3	Consonant
Narrow Tritone (F♯/G♭)	1.414214	7/5	Dissonant
Perfect fifth (G)	1.498307	3/2	Consonant
Minor sixth (G♯/A♭)	1.587401	8/5	Consonant
Major sixth (A)	1.681793	5/3	Consonant
Minor seventh (A♯/B♭)	1.781797	16/9	Dissonant
Major seventh (B)	1.887749	15/8	Dissonant
Octave (C)	2	2/1	Consonant

Table 12.2   Results of Euler's function for a few important musical intervals

Intervals	Euler function values
Unison	$\Gamma(1) = 1$
Octave	$\Gamma(1/2) = 2$
Perfect fifth	$\Gamma(2/3) = 4$
Perfect fourth	$\Gamma(3/4) = 5$
Major third	$\Gamma(4/5) = 7$
Minor third	$\Gamma(5/6) = 8$
Major second	$\Gamma(9/10) = 10$
Minor second	$\Gamma(15/16) = 11$
Tritone	$\Gamma(32/45) = 14$

relation between integer ratios and dissonance. Let n be a positive integer and suppose that its prime factorization is given by [Sándor, 2009]

$$n = p_1^{a_1} p_2^{a_2} \cdot p_N^{a_N}, \tag{12.3}$$

in which the p_k are distinct primes, and $a_k \geq 1$. Then,

$$\Gamma(n) = 1 + \sum_{k=1}^{N} a_k(p_k - 1), \tag{12.4}$$

in which $\Gamma(n/m) = \Gamma(n \cdot m)$, for m integer, was defined as the *Gradus-suavitalis function*, by Euler, and $\Gamma(1) = 1$ by definition. Table 12.2 presents the results obtained for a few important ratios of musical intervals.

These numbers are, according to Euler, a measure of the dissonance of an interval. The larger the result, the more unpleasant the interval would be, which is usually in accordance with the Western listening convention.

13

Note, Time, and Frequency

"If I had my life to live over again, I would have made a rule to read some poetry and listen to some music at least once every week."

Charles Darwin

13.1 Tempo and Time in Music

This chapter deals with some metrics in music, and includes the definition of *tempo* and note, and the representation of a note in time and frequency, as a paradigm for the staff.

Tempo is the overall speed or pace of the musical beat, and it is usually measured in beats per minute (BPM) [Levitin, 2006]. The main function of the beat is the coordination of synchronized movement, such as a dance, and the provision of a common temporal reference for group performance [Patel, 2010].

Of course, the beat has also been used to synchronize exercises in a gym, or to coordinate the movement of troops, in a battle, or to synchronize the athletic movements in a gymnastic competition. The common individual has a remarkable memory for *tempo*, and can detect, on average, a small variation of the parameter, such as, 4% in *tempo*.

The cerebellum is probably the responsible for this accuracy, because it is believed to keep track of the time during the daily lives of the people. It is able to remember the settings used to synchronization to a certain music, as it is being heard, and can recall those settings when the song is sung from memory [Levitin, 2006].

It is difficult to follow a beat that has a period smaller than 200 ms, or larger than 1.2 s. The brain has a preference for beats in the range from 500 ms to 700 ms. The average duration of temporal intervals between stressed syllables is close to, or within, this range [Patel, 2010].

The syncopation is an off-the-beat accent, a temporary displacement of the regular metrical accent in music that is typically caused by stressing the weak beat. The rhythmic accents are placed on weak portions of the beats. Syncopated eighth notes, for example, emphasize the upbeat and form an important part of rock, blues, funk rhythm and blues (R&B), soul, Latin jazz, and country music [Friedland, 2004]. Syncopation is a way to add interest to music, by placing emphasis on beats that would normally be less important [Powell, 2010].

The beats are organized into recognizable accent patterns (2/4, 3/4, 4/4, etc.), and the way in which the beats, or pulses, are grouped together is referred to as meter [Levitin, 2006]. In this regard, a measure is a rhythmic grouping, set off in written music by a vertical barline.

13.2 A Definition of Note

A note, in music notation, is a black or white oval-shaped symbol, with or without a stem or flag, that represents a specific rhythmic duration or pitch. It is also the hearing perception of a unique sound [Juanilla, 2014].

The notes are the basic blocks of the written music. Figure 13.1 shows the notes Do (C), Re (D), Mi (E), Fa (F), Sol (G), La (A), and Si (B) on the staff, using the clefs of Sol (treble) and Fa (bass).

A binary system is used to define the duration of the notes. Each subsequent note has half the duration of the previous one. Figure 13.2 shows the equivalence between the values, or durations, of notes.

Figure 13.1   Notes on the staff, using the clefs of Sol and Fa.

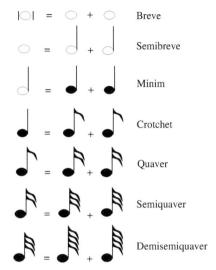

Figure 13.2 Equivalence between the values, or durations, of notes.

| 2 | 1 | 1/2 | 1/4 | 1/8 | 1/16 | 1/32 | 1/64 |

Figure 13.3 Relative values of the notes.

It is called tone in American English, and the usual note values are: breve (double whole note), semibreve (whole note), minim (half-note), crotchet (quarter note), quaver (eighth note), semiquaver (16th note), demisemiquaver (32th note), hemidemisemiquaver (64th note) [Károlyi, 2002].

The double note (breve) is the longest note in the list, the whole note (semibreve) is half the duration of the double note, the half-note (minim) is half the duration of the whole note, the quarter note (crotchet) is half the duration of the half note, the eighth note (quaver) is half the duration of the quarter note, the 16th note (semiquaver) is half the duration of the eight note, and the 32th note (demisemiquaver) is half the duration of the 16th note. Figure 13.3 shows the relative values of the notes.

A note is, actually, a combination of a tone with the time onset and duration, [Loy, 2011a]. This is shown in Figure 13.4, that displays a signal corresponding to a note played on a piano, in which it is possible to identify the envelope, the onset, the duration, and the peak amplitude A.

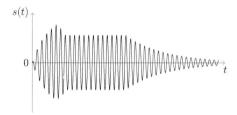

Figure 13.4   A signal corresponding to a note played on a piano.

More precisely, a tone is a sound played or sung at a specific pitch. The tone also involves a pitch, a certain intensity, and the instrument or voice timbre. The combination of a tone with the time onset and duration produces a note [Loy, 2011a]. Diminution is the action of shortening the note values of a theme, usually to render it twice as fast.

The *fermata* is an Italian term that means to stay, indicates a pause when placed on a note or a rest. It is a symbol that indicates that the note should be prolonged beyond the normal duration of a note, usually double the value [de Lacerda, 1966].

Because of the similarity with the wolf howl at night, a wolf note, on any type of acoustic instrument, is markedly different in tone or quality from the others. It is a note that sounds weak or irregular because of the properties of acoustic resonance.

13.3 Time and Frequency Representations of a Note

The staff is a convenient way to represent the most important properties of the tone, its time duration, and its frequency, or pitch. But it is possible to present those features in a Fourier spectrum, associated with a timeline representation.

Figure 13.5 illustrates the relation between the type and position of the notes on the staff, the associated pitches, p_n, time durations, Δt_n, and frequencies, f_n, for perfect, or pure, tones. Sinusoidal signals approximate the sound produced by a flute, or by a whistle, for example.

The amplitude function, $A(\omega)$, represents the product of the loudness by the time duration of the tone, therefore, the *breve*, which has a long duration, is represented by an impulse of higher amplitude. Each note has an associated bandwidth, Δf_n, which depends on the partials that are added to the pitch.

Evidently, a pure tone has no partials, and its bandwidth is zero. That is the reason a perfect tone is usually represented by an impulse, a generalized

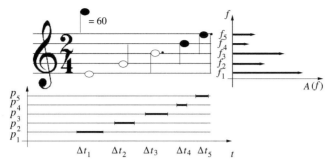

Figure 13.5 Relation between the time and frequency in music.

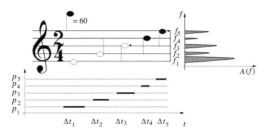

Figure 13.6 Relation between the time and frequency for a piano.

function in which all energy is concentrated in a single frequency. This can only happen if the tone spreads in time, from minus infinity $(-\infty)$ to plus infinity $(+\infty)$, therefore, a pure tone is an idealization.

Because of the harmonics produced, mostly by the attack and decay part of the envelope, Figure 13.6 is a more realistic illustration for the case of a piano. The harmonics enlarge the sound bandwidth and produce the specific timbre of the instrument.

If a tone stretches in time, therefore increasing its duration, it shrinks in terms of frequency, thus decreasing its bandwidth, and vice versa. This is the result of Fourier scaling property. It is common to speak of the product $\Delta t \times \Delta f$ as the time-frequency uncertainty relation. The time-frequency product attains a minimum for the case of a Gaussian envelope, which gives $\Delta t \times \Delta f = 2\pi$.

The energy of the note, E, is given in joule [J], and it usually increases if the note is elongated in time T. For example, the energy for a given signal $x(t)$, in the interval $[0, T]$ is defined as

$$E = \int_0^T |x(t)|^2 dt. \tag{13.1}$$

If one considers the envelope signal

$$x(t) = e^{-t}u(t), \tag{13.2}$$

in which $u(t)$ is the unit step, or Heaviside, function, then

$$E = \int_0^T |e^{-t}u(t)|^2 dt = \int_0^T e^{-2t} dt = \left[-\frac{e^{-2t}}{2}\right]_0^T = \frac{1}{2}\left[1 - e^{-2T}\right] \text{ J.} \tag{13.3}$$

The result increases, as $T \to \infty$, and reaches a maximum value of $\frac{1}{2}$ J.

The energy is generally defined for the whole signal domain, as

$$E = \int_{-\infty}^{\infty} |x(t)|^2 dt. \tag{13.4}$$

If one recalls Fourier transform properties, it is possible to write the energy using both the time and frequency domains,

$$E = \int_{-\infty}^{\infty} |x(t)|^2 d = \frac{1}{2\pi} \int_{-\infty}^{\infty} |X(\omega)|^2 d\omega, \tag{13.5}$$

in which $X(\omega)$ is Fourier transform of $x(t)$.

The timbre is another term for tone color, it distinguishes one instrument from the other, when both are playing the same written note. The tonal color is produced by overtones from the instrument vibratory movement [Levitin, 2006].

As can be seen in the figure, a *dotted note* is a written note with a dot to the right of it. The dot adds half the rhythmic duration to the note's original value. The indication for the value of the quarter note (60), above the staff, refers to the repetition time of a note, or *tempo*. In this case, the crotchet is repeated every second, or 60 times per minute. It is a useful indication for the metronome.

Figure 13.7 shows the staff with the clef of Sol (G), the beat marking indicating two beats per measure (2) for the quarter note (4), and the measure bars. The *tempo* is set to 80 beats per minute.

An octave is a musical interval between two pitches in which the upper pitch vibrates twice as fast as the lower. The interval is equivalent to 12 semitones, in a chromatic scale, or eight tones, in a diatonic scale.

If a reference tone has frequency f_0, then the frequency of the nth octave of this reference can be computed by the formula

$$f_n = f_0 \cdot 2^n, \quad n = 1, 2, 3, \ldots. \tag{13.6}$$

Figure 13.7 Staff, clef, beat and measure.

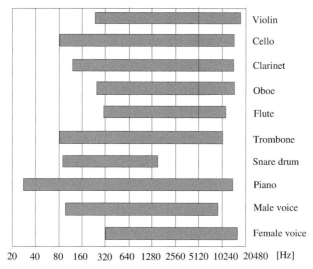

20   40   80   160   320   640   1280   2560   5120   10240   20480   [Hz]

Figure 13.8 The approximate tonal range of some instruments.

Therefore, the frequency two octaves above, or two octaves up, the note La (A) is

$$f_2 = 440 \cdot 2^2 = 1,960 \text{ Hz.}$$

The range is the distance between the lowest and highest possible notes of an instrument or melody. Figure 13.8 illustrates the approximate tonal range of some instruments.

The rhythm is an important element of music as it unfolds in time. Rhythm is commonly described as the component of music that punctuates time, that gives the placement of a note in time [Wyatt and Schroder, 1998]. Rhythmic structures are organized into *measures*, which indicate the division of time into groups of beats [Martineau, 2008].

It can be perceived that rhythm is also the ordering of the sound, or noise, and silence in time, following a certain pattern of periodicity, or cyclostationarity [Abad, 2006].

14

Theory of Chord Formation

"The music is not in the notes, but in the silence between."

Wolfgang Amadeus Mozart

14.1 Chord Formation

The harmony draws its tones from major or minor scales to produce chords, vertical structures of three or more notes that sound together. Chords are identified by two symbols, a letter that represents its root, and another one that specifies the type of chord. The common qualities are major, minor, augmented, and diminished. [Franceschina, 2015].

The chord is a result of some rules, a harmonic combination that has three or more pitches sounding simultaneously. There are books that display set of charts including the usual triads, 6ths, 7ths, 9th, 11th, and 13th chords, using a common cipher system. They generally use the English notation C, D, E, F, G, A, and B for the seven notes of the diatonic scale, corresponding to seven diatonic tones of C major: *Do, Re, Mi, Fa, Sol, La,* and *Si* [Chediak, 1984].

Figure 14.1 shows the major chords, and Figure 14.2 shows the minor chords for a piano.

14.2 Triad

A *triad* is the most basic chord structure, a three-note chord built on alternating scales steps. Usually, a triad encompasses the first, third, and fifth notes of a scale. The triad is the basis for most popular music, including rock, dance music, reggae, and samba.

The major triad, shown in Figure 14.3, for the scale in C, occurs naturally in the harmonic series, and is the foundation of the tertial music, the chords built in thirds, for example C-E-G, which corresponds to the major chord Do-Mi-Sol. The movement of a note, from the bottom to the top of a triad,

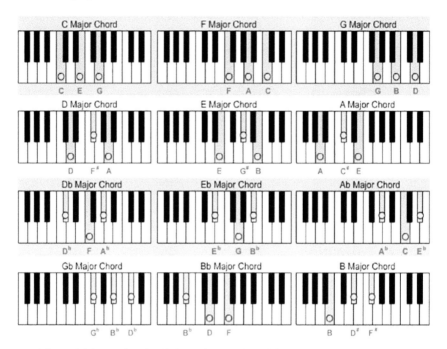

Figure 14.1   Major chords for a piano. Adapted from https://br.pinterest.com.

produces an inversion, that retains the same notes, but changes the bass [Martineau, 2008].

The major triads in first inversion have two minor intervals, giving a nostalgic feeling. On the other hand, if a minor triad is inverted it sounds typically major, because two intervals are major. Diminished or augmented chords are considered rootless, because they do not present a stable fifth or fourth.

The minor triad is obtained by lowering the second note of the major chord by a semitone, for example C-E♭-G, which corresponds to the major chord sequence Do-Mi♭-Sol. which implies dividing the frequency of the second note by $\sqrt[12]{2} = 1.05946$, for the tempered chromatic scale. The minor chord in C is shown in Figure 14.4.

The augmented triad consists of a major third on the bottom and a minor third on top. It is a volatile chord that seeks upward resolution. For example C-E-G♯, which corresponds to the major chord Do-Mi-Sol♯.

The diminished triad consists of a minor third on the bottom and a minor third on top, for example C-E♭-G♭, which corresponds to the major chord

Figure 14.2 Minor chords for a piano. Adapted from https://br.pinterest.com.

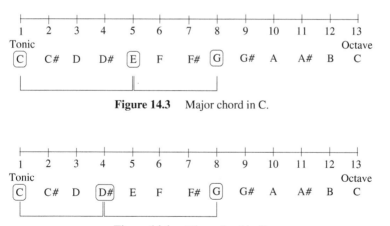

Figure 14.3 Major chord in C.

Figure 14.4 Minor chord in C.

Do-Mi♭-Sol♭. It is an unstable chord that seeks resolution downward, used to add some drama to the piece, because it is the least sturdy and assured of the triads [Franceschina, 2015].

14.3 Tetrachord

A *tetrachord* is a four-note chord, that corresponds to a combination of a whole-step, another whole-step and a half-step. The word is based on the *tetrachordon*, a Greek four-stringed harp, also called *lyre*. The four tones encompassed by the musical instrument composed a perfect fourth (4th), and used to be called a *tetrachord*, the primordial component that later became the basis of the current musical scales.

If a tetrachord is played beginning on C, the notes are C, D, E, and F. The combination of the previous tetrachord with the one that begins on G, that is G, A, B, and C, produce an eight-note scale in the Ionian mode, currently known as a natural major scale, that is C, D, E, F, G, A, B, and C.

If the same tones are used, beginning on the sixth (6th) note of the combined two tetrachords, one obtains the notes A, B, C, D, E, F, G, B, and A, which constitute the Aeolian mode, also known as the natural minor scale [Palmer et al., 1994].

Diabolus in Musica is the name given to the tritone, an interval of three whole steps, in the 17th century. Because it was considered dissonant, the use of the interval was prohibited by music theorists of the time.

15

Study of Melody and Composition

"Music gives a soul to the universe, wings to the mind, flight to the imagination and life to everything."

Plato

15.1 Definition of Melody

Melody is the musical element that deals with the horizontal presentation of pitch, it is a series of sounds with different intensities, pitches, and durations, that succeed each other in time, with a musical sense [Abad, 2006].

It is said that the melody is the more abstract element of music. Different from the rhythm, which is linked to the body, and the harmony, which is related to the mind, the melody seems to evolve from the artist's poetic inspiration. It is also a result of the artist's heritage, culture, and mother language. A melody is composed of strong and ornamental notes [Juanilla, 2014].

15.2 Elements of a Melody

The motif, or motive, is the fundamental element of the musical composition, and consists of a short melodic design formed by two or more notes. It is considered the seminal musical idea, and will repeat itself along the musical piece.

The semi-phase, or sub-phrase, consists of two or three motives, or can be a long motive with a stronger than usual closure. The first semi-phrase is known as antecedent, a kind of proposal, a question or a call for the next semi-phrase, the consequent or contrasting idea, that replies as a following part. Two or three semi-phrases form a phrase.

The theme is the melodic idea used as the basis of development. It is also the main self-contained melody of a musical composition . After the theme is stated, then it undergoes a series of sectional alterations.

A phrase is a small musical unit, a sub-section of a melody, equivalent to a grammatical phrase in a sentence, a substantial musical thought, which ends with a musical punctuation known as *cadence*. A *cadence* is a melodic or harmonic punctuation mark at the end of a phrase, major section or entire work. The phrase is a unit of musical meter that carries a complete musical sense, built from figures, motifs, and cells [Juanilla, 2014].

The cadence is a sort of musical punctuation, that varies in conclusiveness, such as, the comma, the semi-colon, the colon, and the period in language. The usual cadences are classified as: authentic, perfect authentic, semi, plagal, and deceptive.

Musical periods are similar to sentences in language. They have usually eight measure themes, split between two phrases, and they contain an antecedent and a consequent phrase.

There are different types of melody:

Retrograde – A melody presented in backward motion.

Retrograde inversion – A melody presented backwards and intervalically upside down.

Inversion – A variation technique, in which the intervals of a melody are reversed. Table 15.1 shows the transformations that occur when the intervals are inverted [Abromont and de Montalem bert, 2010]. To find the name of the inverted interval just subtract the interval number from nine.

Table 15.1   How an interval is transformed after inversion

Intervals	After inversion
Unison	Octave
Second	Seventh
Third	Sixth
Fourth	Fifth
Fifth	Fourth
Sixth	Third
Seventh	Second
Octave	Unison

Table 15.2   Qualification of inverted intervals

Interval qualities	After inversion
Minor	Major
Major	Minor
Perfect	Perfect
Diminished	Augmented
Augmented	Diminished

On the other hand, the qualification of an inverted interval is always the opposite of the qualification of the original interval, as shown in Table 15.2.

Modulation is the process of changing from one musical key to another. It is a change of tonality during the composition. For instance, if a piece begins and ends in Do major (C major), a tone given by the tonic, and it changes to Sol major (G major), then there is a modulation to the dominant tone, the fifth note above the tonic [Bennett, 1998].

15.3 Functional Harmony

The tonal era appeared at the end of the Renaissance, as an improvement of the modal practice, that was adopted during the Middle Ages. The theory of harmony evolved from the daily practice of the musicians, in that period, based on the melodic lines of polyphonic pieces. The pleasantly sounding lines started to be treated as formulas for tonal cadences, and finally resulted in perfect, major or minor chords.

Therefore, harmony is the very matter of music. The study of harmony involves the construction of chords, the development of chord progressions, and the connections between them. The concept of interval is important to harmony, because it is considered a building unit of the chord [Almada, 2009]. There are three basic levels to understand harmony:

Sonority – It is how the notes sound together to make harmony, or the quality of having a profound and pleasant sound.

Harmonic relation – It is the relationship between certain chords within a key. In contrast, non-harmonic relation is a type of dissonance that occurs in polyphonic music.

Functional harmony – It is how the chords relate to form of the composition and the overall tonal scheme. It is a theory of tonal music that considers all harmonies to function as tonic, dominant, or subdominant harmony.

Harmony takes many forms, and can elevate a musical piece from conventional and predictable to refined and provocative. In this context, the common forms of harmony are:

Diatonic harmony – This kind of harmony refers to a type of music in which the notes and chords are linked to a master scale. Diatonic harmony comes from ancient Greek instrumentals, through Renaissance chorales to contemporary music.

Non-diatonic harmony – This type of harmony introduces notes that are not all part of the same master scale. This harmony forms the basis of jazz.

Atonal harmony – A harmony in which the music lacks a tonal center, or key. It is not built on a major or minor scale. In fact, atonality is the absence of functional harmony.

15.4 Fundamentals of Composition

Musical composition is the creation of an original piece of music, either vocal or instrumental, to be executed in a repetitive way, as opposed to improvisation music, in which each performance is unique. The music can be preserved in the memory, or by a writing or notation system [Howard, 1991].

Composition is the process of making or forming a piece of music by combining, in a creative way, the parts, or elements of music. It is also a manner of organizing sounds over time, and this activity involves complex cognitive processes [Wade-Matthews and Thompson, 2003].

It is also the structure of a musical piece, or the process of writing a new piece of music. Composition is one of the manners to be creative in music, but there are others, such as, interpretation [Howard, 1991].

A good composition has form and meaning, it tells a story, or invokes a mood. It is possible to create a chord progression and then fit a melody to the chords, as in harmonic composition; or to create a melody and then place the chords, as in melodic composition.

Holistic composition is a manner of creation in which the composer progresses measure by measure, producing the melody and harmony simultaneously [Miller, 2005].

In this context, it important recall that a melody is a collection of pitched sounds arranged in musical time, in accordance with given cultural conventions and constraints [Patel, 2010].

15.4.1 Musical Parameters

In order to begin the composition process, it is useful to establish the some basic parameters for the musical piece.

1. The tempo – The speed or pace of the musical beat, measured in BPM.
2. The time signature – The number of beats, or pulses, contained in each measure, and which note value is equivalent to a beat.
3. The key signature – The series of sharps or flats written on a musical staff to indicate the key of a composition.
4. The instrument – Any object that produces sound to play the piece.

For example, it is possible to start the process choosing the following music parameters.

1. Tempo – A tempo of 80 BPM, as shown in Figure 13.7.
2. Time signature – Figure 13.7 also pictures a time signature of 2/4, that means two beats per bar, using the quarter note (crotchet).
3. Key signature – Do (C) major is a very common key signature to begin with.
4. Instrument – The musician can pick his or her own favorite instrument, such as, a piano or a guitar.

After that, it is necessary to define parameters related to dynamics, range, rhythm, and duration.

1. Dynamics – The varying degrees of volume in the performance of music, such as, loud of soft.
2. Range – The distance between the lowest and highest possible notes of an instrument or melody.
3. Rhythm – The combinations of long and short duration sounds, that convey a sense of movement and regularity in time.
4. Duration – That is, the length of time a pitch, or tone, is sounded.

For instance, the composer has several degrees of freedom to choose those musical parameters.

1. Dynamics – It is very common to start the piece very soft, or *pianissimo*, that means a dynamic marking (*pp*), and then increase the loudness to a certain level, such as *mezzo-piano*, a medium quiet dynamic marking (*f*).
2. Range – The music can start in low Sol (G), for example, and increase in pitch to explore the vicinity of Do (C).
3. Rhythm – A composer can draw from elements of folk music, for example, and include them in his or her composition.
4. Duration – The duration of the notes can be, for instance, three beats long.

As implied by the earlier discussion, music theory is necessary to understand why the music has the emotional effect that it elicits, and it is important to focus on a few essential elements of the music. The composer must be able to read music notation.

1. Melody – The effect of the single line to impress the audience, and how to handle the problems of writing the musical score.
2. Harmony – How notes and lines sound together at the same time.
3. Form – How any section of the piece can sound like a beginning, middle, or end, and therefore how to organize it in unique ways to tell the musical story.

The apparent logic in music comes from the fact that most of the music that is played follows the same rules. These rules are planted in the brain, along the years, and one expects to hear them. These expectations are incorporated into the music by the composers.

15.4.2 Initial Steps to Composition

The primary identifying parts of any composition are the motif, a short block of notes that serves as an identification point for a longer melody, and the theme, a complete melodic phrase that carries the main musical idea of the piece [Miller, 2005].

The previous musical knowledge, either practical or theoretical, is important to develop the composition theme. Every piece has roots, that can go deep into the musical tradition or culture, and also natural ramifications.

The study of music theory gives the composer the necessary tools to deal with harmony, melody, and rhythm in a proper manner. In the same way, composition is a daily task.

Therefore, the composer is usually preoccupied with the historic perspective, and listens to music that is diversified, because this can provide some feedback to the composition process.

Every music has a specific style, such as, *samba*, rock and roll, *bossa nova*, waltz, pop, and classical. The analysis of musical scores is a key to understand the composition structure.

Transposition is important to work on the melody, that is an essential part of the whole compositional process. On the other hand, the study of solfege introduces a collaborative practice that can improve the composition.

A composition goes beyond melody and chord progression, it uses repetition, variation, modulation, and other techniques, and can be a construction for multiple instruments and voices. Structure and form are important in musical composition [Miller, 2005].

16

Musical Instruments

"Lean your body forward slightly to support the guitar against your chest,
for the poetry of the music should resound in your heart."

Andres Segovia

16.1 Instrumentation

Instrumentation is the combination of instruments that a composition is written for. This chapter presents a compilation of musical instruments, along with their main characteristics [Wade-Matthews and Thompson, 2003; Pfeiffer, 2010; Buck, 2014; Hodge, 2006; Shepheard, 2014].

There are six basic types of musical instruments, according to the way the sound is produced [Martineau, 2008]:

Aerophone – A musical instrument that produces sound primarily by the vibration of the air, without the use of strings or membranes, and without the vibration of the instrument itself.

Chordophone – A musical instrument that relies on the vibration of a set of strings, such as, a violin, or a cello.

Electrophone – A musical instrument in which the original sound either is produced by electronic means, or is generated in a conventional manner and electronically amplified. The electronic music synthesizer and the electronic organ are examples.

Idiophone – A musical instrument that creates sound primarily using the vibration of the whole body, without the use of strings or membranes, such as the marimba, or a triangle.

Membraphone – A musical instrument that relies on the vibration of a membrane, such as, a drum.

Metallophone – A musical instrument that consists of tuned metal bars which are struck to make sound. The glockenspiel and vibraphone are metallophones.

The tone color is a unique characteristic sound of a musical instrument or voice. The tone also involves a duration, a certain intensity and the instrument or voice timbre [Franceschina, 2015].

16.2 Instruments of an Orchestra

The following describes the usual instruments of an orchestra, of a band, and some more. Every instrument must first be tuned using a two-pronged steel device, called tunning fork, that is an acoustic resonator which vibrates when struck to give a note of specific pitch. The tuning fork was invented in 1711 by British musician John Shore (c.1662–1752), an English musician who played the trumpet.

The tunning fork produces an almost pure tone, a sine wave, with the vibrational energy concentrated at the fundamental frequency. The particles of air in the vicinity of the device are set into harmonic motion, with the same period, and the period is preserved when the vibration is passed to the ear drum of the listener [Newman, 2000a].

While tuning forks have traditionally been used to tune musical instruments, electronic tuners tend to replaced them. Most instruments are tuned to A4 (La), at a frequency of 440 Hz. An instrument that is not notated at its sounding pitch is called a transposing instrument [Miller, 2005].

The *cello* is the tenor-ranged musical instrument of the modern string family. It is an abbreviation for violoncello. The cello range goes from C2 (65.41 Hz) to B5 (987.77 Hz). [University, 2019]. Figure 16.1 displays a professional cello.

The bass drum is the lowest-sounding non-pitched percussion instrument. A tambourine is a small drum with metal jingles set into the edges. Both the drum head and the jingles are untuned.

Maracas, which originated from Latin America, are rattles, often made from gourds, a type of squash, filled with dried seeds, beads or tiny ball bearings that produce the rattling sound. Maracas are percussion instruments

Figure 16.1 The cello, a stringed tenor-ranged instrument. (Adapted from Pixabay – https://pixabay.com.)

Figure 16.2 A bass drum, the lowest-sounding non-pitched percussion instrument. (Adapted from Creative Commons – By Ludwig & Ludwig. Public Domain.)

that can also be made of wood or plastic, and the sound they make depends on the material they are made of.

A woodwind instrument is an instrument that produces its sound from a column of air vibrating within a multi-holed tube. Figure 16.2 displays a bass drum.

The double bass is the lowest-sounding instrument of the modern string family. The double bass range goes from E1 (41.20 Hz) to B3 (246.94 Hz). A *sousaphone* is an ultra bass instrument designed for use in marching bands.

Figure 16.3   The contrabassoon is a double-reed instrument. (Adapted from Creative Commons: https://commons.wikimedia.org/wiki/File: Contrabassoon2.png.)

Figure 16.4   The clarinet, a single-reed instrument. (Adapted from Pixabay – https://pixabay.com.)

The *bassoon* is a low-sounding regular instrument of the woodwind family. It is a double-reed instrument. The *contrabassoon*, also known as the double bassoon, is a larger version of the bassoon, sounding an octave lower. It is the lowest-sounding double-reed instrument of the woodwind family cool jazz. It is also a relaxed style of modern jazz, promoted in the 1950s and 1960s by Brubeck and others [Stokes, 2019]. Figure 16.3 displays a contrabassoon.

The clarinet is the tenor-ranged instrument of the woodwind family. It is a single-reed musical instrument. It is a cylindrical bore instrument closed at one end, therefore, the normal resonant modes must have a pressure maximum at the closed end (the mouthpiece), and a pressure minimum near the bell. These conditions result in the suppression of even harmonics in the sound [Roederer, 1998]. Figure 16.4 displays a clarinet.

In addition to the fundamental note, the clarinet plays the harmonics that are produced by odd fractions, such as, $\frac{1}{3}, \frac{1}{5}, \frac{1}{7}, \frac{1}{9}, \ldots$, which means that only the odd overtones are heard, as shown in Figure 16.5. [du Sautoy, 2007].

The *flute* is a metal tubular instrument that is the soprano instrument of the standard woodwind family. The concert flute range goes from C4 (261.63 Hz) to B6 (1,975.53 Hz).

An acoustic guitar is a six-stringed fretted instrument, composed of a body with a soundboard, a head with tuning pegs, a neck with frets and a fingerboard, a bridge, a saddle, and the fixing pegs. There are various types of guitar, in addition to the acoustic one, including the electric, the folk, the

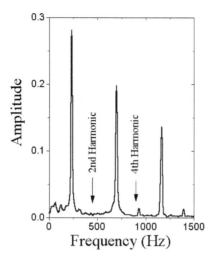

Figure 16.5   The spectrum of the clarinet. (Adapted from Physical Music-Notes, pages.mtu.edu/ suits/clarinet.html.)

electroacoustic, the bass, the *ukulele*, and the *cavaquinho* [Moon, 2015]. Bout is a term applied to the upper and lower sections of the guitar body.

A fret is a small metal strip inserted into the fingerboard of some instruments, such as guitars, to make pitch definition more precise [Pen, 1992]. Camber is the curvature of the fingerboard, also known as the radius.

The air in a guitar's cavity resonates with the vibrational modes of the string and soundboard. At low frequencies the chamber acts like a Helmholtz resonator, named after the German philosopher, physician, and physicist Hermann Ludwig Ferdinand von Helmholtz (1821–1894), who made significant contributions in several scientific fields, including physiology, psychology and physics, in which he is known for theories on electrodynamics and thermodynamics.

Figure 16.6 displays an acoustic guitar, a versatile instrument, used in popular music performance. In standard tuning the guitar's six strings are tuned, from low to high, as E2 (Mi), A2 (La), D3 (Re), G3 (Sol), B3 (Si), and E4 (Mi). The acoustic guitar range goes from E2 (82.41 Hz) to F6 (1396.91 Hz). Appendix C displays the corresponding frequencies and wavelengths.

The soundboard, or top, increases the surface of the vibrating area, in a mechanical impedance matching process, to propagate the sound. A guitar has various sound coupling modes, including the coupling between the

Figure 16.6   The acoustic guitar is a six-stringed fretted instrument. (Adapted from Pixabay – https://pixabay.com.)

Figure 16.7   The keyboard is any instrument whose sound is initiated by pressing keys. (Adapted from Pixabay – https://pixabay.com.)

string and the soundboard, the soundboard and the cavity air, and both the soundboard and the cavity air to the outside air.

The *harp* is a plucked instrument having strings stretched on a triangular frame. Harps have been in use since antiquity in Europe, Middle East, Asia and Africa, dating back at least as early as 3000 BC. The modern harp has 47 strings, tuned to the equal-tempered scale [de la Fuente, 2014].

A keyboard instrument is any instrument whose sound is initiated by pressing a set of keys with the fingers. The piano, the harpsichord, the organ, and the synthesizer are the most common types. Figure 16.7 displays a general use keyboard.

The *harpsichord* was an ancient keyboard instrument, whose sound was produced by a system of levered picks that pluck its metal strings. It was a common musical instrument in the Renaissance and Baroque eras.

The *oboe* is nasal-sounding double-reed instrument that is the alto of the standard woodwind family. The oboe produces mainly harmonics equivalent to the fourth, fifth, and sixth partials, and others at higher pitches, as the 10th partial. Therefore its sound is considered as more brilliant, vivid. The *English horn* is a tenor oboe, and also a richly nasal-sounding double-reed woodwind instrument.

The *organ* is a wind or keyboard instrument, usually with many sets of pipes controlled from two or more manuals, including a set of pedals played by the organist's feet. Also, a set of mechanical or electrical stops allow the player to open or close the flow of air to selected groups of pipes. The organ was devised in Alexandria in the 3rd century B.C., and became one of the main instruments of Rome by the 2nd century AD [Wade-Matthews and Thompson, 2003].

The *piano* is a versatile modern keyboard instrument that makes sound via keys that connect to felt-tipped hammers to strike a set of internal strings.

The *pianoforte*, or fortepiano, was the original instrumental prototype of the piano, which appeared in the late Baroque or early Classic eras. Piano is also a term used in music dynamics to indicate a soft way of playing [Shepheard, 2014].

The ancestor of all the keyboard musical instruments, including the piano, was the organ, invented around 300 B.C by Cresibius of Alexandria, who named it *hydraulis*. The instrument applied a mechanical wind supply to a set of giant panpipes.

The direct ancestor of the piano was the *gravicenmbolo col piano e forte*, which can be translated to harpsichord with soft and loud, constructed by Bartolomeo Cristofori, in 1709. [Burrows, 2004]. It was originally named pianoforte because it could be played softly or loudly, according to the force applied to the keys [Henry, 2003].

A standard piano keyboard has 88 keys, and goes from A0, with a frequency of 27.50 Hz, to C8, with a frequency of 4186.01 Hz. What is referred to as middle C is C4, with a frequency of 261.63 Hz. Figure 16.8 displays a modern piano.

A prepared piano is a modern technique, invented by composer John Cage, in which various natural objects, including spoons, erasers, or screws, for example, are inserted between the strings of a piano, in order to create unusual sounds.

The *reed* is a flexible strip of cane, or metal, that vibrates in the mouthpiece of a wind instrument. The *snare drum* is a non-pitched drum with two heads stretched over a metal shell. The lower head has metal wires strapped across it to produce a rattling sound.

Timpani are various-sized kettle-shaped pitched drums. Also, a tenor instrument of the percussion family. It is similar to kettledrums, and are seldom used alone. Every orchestra has two or more, a small one for the

Figure 16.8   The *piano* is a keyboard instrument. (Adapted from Pixabay – https://pixabay.com.)

high pitches and a large one for the low pitches. The required pitch is written on the clef of Fa (F) [Károlyi, 2002]. The *tabla* is a pair of drums used to accompany the music of India.

A brass instrument is a powerful metallic instrument with a mouthpiece and tubing that must be blown into by the player, such as trumpet, trombone, French horn, tuba, baritone, and bugel. The *tuba* is a large valved brass instrument, the bass of the modern brass family. The tuba range goes from F1 (43.65 Hz) to F4 (349.23 Hz).

Saxophone, or sax, is a family of woodwind musical instruments with a single reed and brass body. It is commonly used in jazz and marching bands, or concert ensemble music. Figure 16.9 displays a saxophone.

The trombone names a family of brass instruments that change pitch via a movable slide (alto, tenor, and bass versions are common). It is a unique instrument, because it appeared, in its current form, in the 15th century, and no important modification has been made ever since. Its notation is written on the clef of Fa (F).

The trombone sound is powerful, and somewhat similar to the trumpet. The tenor trombone is tuned in Si♭ (B♭), and the bass trombone is tuned in

Figure 16.9   The *saxophone* is a woodwind instrument with a single reed. (Adapted from Pixabay – https://pixabay.com.)

Sol (G). The *sackbut* is an ancient brass instrument, ancestor to the trombone [Károlyi, 2002]. Figure 16.10 displays a typical trombone.

The *trumpet* is a valved instrument that is the soprano of the modern brass family. Figure 16.11 shows a typical trumpet.

A trumpet mouthpiece produces a pulse train, which carries more spectral energy at higher frequencies, that is filtered by the instrument, to generate the trumpet characteristic sound. The *cornet* is a mellow-sounding member of the trumpet family. Figure 16.12 shows the trumpet spectrum [Doering, 2012].

The *French horn* is a valved brass instrument of medium to medium-low range, that is alto to bass. It is usually tuned in the clef of Fa (F). [Károlyi, 2002]. Is also called simply horn. The main harmonics generated by the French horn are of lower pitches, mainly the first, second, third, and fourth partials. Its sound is considered dull, in a certain sense [Abad, 2006]. Figure 16.13 shows a french horn.

The *xylophone* is a pitched percussion instrument consisting of flat wooden bars on a metal frame that are struck by hard mallets. The xylophone originally came from Africa or Asia, and has a Greek name that means wood sound. The *glockenspiel* is a pitched-percussion instrument comprised of metal bars in a frame struck by a mallet that looks like a miniature xylophone.

A *marimba* is a pitched percussion instrument comprised of wooden bars struck by mallets. It is a softer and mellower version of the xylophone. Figure 16.14 displays a marimba.

According to Yamaha's Musical Instrument Guide, the xylophone and the marimba look identical, but the marimba is tuned on even-numbered harmonics, with tuning on the fundamental pitch, the fourth harmonic, and

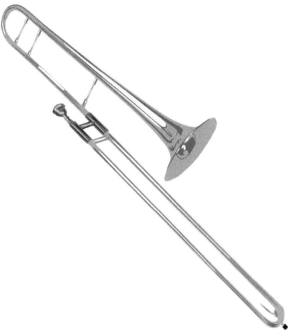

Figure 16.10   The trombone is an instrument from the XV century. (Adapted from Pixabay – https://pixabay.com.)

Figure 16.11   The trumpet is a valved instrument. (Adapted from Pixabay – https://pixabay.com.)

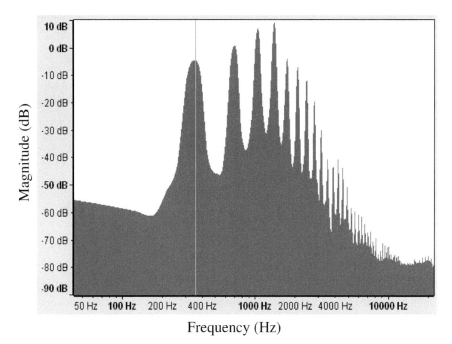

Frequency (Hz)

Figure 16.12   The trumpet spectrum. (Adapted from copyrighted material by Ed Doering, Creative Commons Attribution 2.0 license (creativecommons.org/licenses/by/2.0/).)

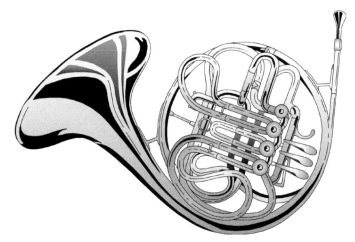

Figure 16.13   The french-horn is metallic instrument tuned in Fa (F). (Adapted from the Open Clip Art Library, as public domain – openclipart.org.)

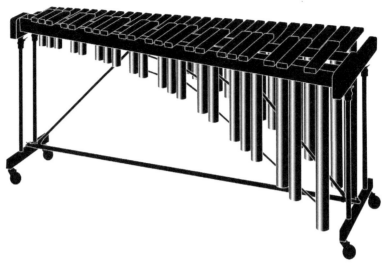

Figure 16.14   The marimba is a pitched percussion instrument. (Adapted from Pixabay – https://pixabay.com.)

the 10th harmonic. The xylophone, however, is tuned on the fundamental pitch and the odd-numbered third harmonic [Yamaha, 2019].

Tubular bells are musical instruments in the percussion family. Their sound resembles that of church bells, carillon, or a bell tower. Also called chimes.

The *violin* is the soprano instrument of the modern string family. Is has a very rich sound. In addition to the fundamental note, the violin produces all the harmonics devised by Pythagoras, such as, $\frac{1}{2}, \frac{1}{3}, \frac{1}{4}, \frac{1}{5}, \ldots$. This is the famous harmonic series. [du Sautoy, 2007]. Figure 16.15 displays a violin.

A *string instrument* is a musical instrument that is played by placing the hands directly on the strings, such as violin, viola, cello, double bass, harp, guitar, dulcimer, psaltery, and the ancient viols. *Violoncello* is the full name of the cello. It is the tenor instrument of the modern string family.

Viola is the alto instrument of the modern string family. It is usually played with a bow [de Lacerda, 1966].

16.3 Instruments of a Rock Band

An electric instrument has its sound produced or modified by an electromagnetic pickup, while an electronic instrument has its sound produced or modified by an electronic device.

Figure 16.15   The violin is the soprano instrument of the string family. (Adapted from Pixabay – https://pixabay.com.)

The electric guitar is a six-stringed fretted instrument that uses sensors (pickups) to convert the vibration of the strings into electrical signals. The strings vibrate when the guitar player strums, plucks, finger-picks, slaps, or taps them. The sensor generally uses electromagnetic induction to create this signal. The signal is fed into a guitar amplifier and sent to the speakers, to be converted into an audible sound. Figure 16.16 displays an electronic keyboard.

The electric bass guitar, also known as electric bass or simply bass, is the analogous of a *double bass* for acoustic music, it is the lowest-sounding string instrument in a band, and the strings are numbered as G (1st), D (2nd), A (3rd), and E (4th). The bass clef is commonly used to play the bass guitar [Friedland, 2004].

The sound of each bass strings is tuned an equal distance from the string above it, therefore, the instrument is symmetrical. This means that if a scale is played starting on one string, the same fingering can be used to play the same scale on a different string [Pfeiffer, 2010].

The bass has a lower tune than the guitar, and the deep notes of the bass are on the lower end of the sound spectrum. The four-string bass tune is equivalent to the double bass, which corresponds to pitches one octave lower

Figure 16.16   The electric guitar is the symbol of a rock band. (Adapted from Pixabay – https://pixabay.com.)

Figure 16.17   The electric bass guitar, an instrument invented by Paul Tutmarc from Seattle, Washington. (Adapted from Creative Commons.)

than the four lowest-pitched strings of a guitar. The four-stringed bass guitar range goes from E1 (41.20 Hz) to C4 (261.63 Hz). Figure 16.17 shows an electric bass guitar.

A bass guitar is slightly longer than a regular electric guitar, and their strings are thicker. Therefore it has a deeper sound, one octave lower than the guitar strings. The strings are usually tuned to E, A, D, and G, beginning with the fourth open string [Capone, 2009].

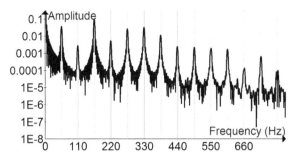

Figure 16.18 The spectrum of the bass guitar. (Adapted from Wikipedia, Creative Commons Attribution-ShareAlike License – en.wikipedia.org/wiki/Harmonic_analysis.)

Figure 16.19 The electronic keyboard, a versatile instrument. (Adapted from Creative Commons.)

Figure 16.18 shows the spectrum of an electric bass, for the open string La (A). Note that there is a peak at 55 Hz, the fundamental frequency or tone, but the spectrum contains peaks at 110 Hz, 165 Hz, and at other frequencies, the harmonics, corresponding to multiples of the fundamental frequency.

An electronic keyboard, also called digital keyboard, is a musical instrument that includes a synthesizer, an amplifier, and a set of loudspeakers. It can produce a wide range sounds, and emulate other musical instruments, such as, a piano, an electric piano, an organ, a pipe organ, or a violin. Figure 16.19 displays an electronic keyboard.

A drum set, also called a drum kit, combines a snare drum, a bass drum, and cymbals. A standard kit contains a snare drum, mounted on a stand, which may include rutes or brushes; a bass drum, played by a pedal; one or more toms, played with sticks or brushes; two cymbals mounted on a stand, a hi-hat, played with the sticks, that are opened and closed with a pedal; and

Figure 16.20   A drum set. (Adapted from Creative Commons.)

one or more cymbals, mounted on stands, played with the sticks. Figure 16.20 displays a drum set.

16.4 Special Instruments

The *berimbau* is a kind of monochord, a single string percussion instrument, with African origins. It is the main instrument used to produce the complex rhythms in Brazilian music that accompanies *capoeira*, a Brazilian martial art.

It consists of a flexible wooden bow called *biriba* or *verga*, that exerts a tension on a steel string, called *arame*. A gourd called *cabaça* ressonates to amplify the sound, and produces the characteristc *timbre* of the instrument.

The *berimbau* is played with the help of a small, thin stick called *baqueta* or *vareta*. A metal or stone disk, called *dobrão* or *pedra*, is used to adjust the *pitch* during the performance, and a shaker, called *caxixi*, is used as a percussion device. A picture of a professional *berimbau* can be seen in Figure 16.21 [Lyra, 2019].

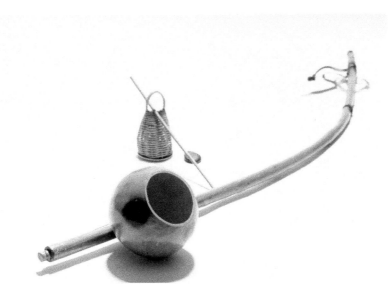

Figure 16.21 The berimbau is a monochord instrument used to accompany the Brazilian capoeira. (Adapted from Lyra, 2019.)

The *zabumba*, shown in Figure 16.22, is a type of bass drum, a compressed double drum, used in Brazilian folk music, such as *baião*, *forró*, *xaxado*, and *xote*. The player wears the drum, using a strap, while standing up, and uses both hands to play on both sides of the *zabumba*.

The superior head, the low tuned side of the *zabumba*, is played with a padded mallet. It is the heartbeat of the *forró* and *baião* rhythms. The resonance head is played with the *bacalhau* (a thin switch), to produce the typical contrast with contra-rhythms, the call and response rhythmic pattern.

Viol is an ancient string instrument, that is an ancestor to the modern violin. *Viol' da gamba* is a Renaissance bowed string instrument held between the legs like a modern cello.

The *koto* is a Japanese plucked instrument with 13 strings and movable bridges. The *lute* was an ancient pear-shaped plucked instrument widely used in the Renaissance and Baroque eras. The *shamisen* is a banjo-like Japanese stringed instrument.

Ud is a lute-like, pear-shaped, fretless stringed instrument commonly used in music from the Middle East. *Shakuhachi* is a Japanese flute. The *Shawm* is an ancient double-reed woodwind instrument.

Figure 16.22   Zabumba is a type of bass drum. (Adapted from Creative Commons – Public domain image.)

The *cymbals* is a percussion instrument, usually consisting of two circular brass plates struck together as a pair. The *gong* is a non-pitched percussion instrument made of a large metal plate struck, also known as the *tamtam*. It is a very large metal plate that is suspended from a metal pipe. It is similar to a cymbal and is also untuned. The *chimes* is a percussion instrument comprised of several tube-shaped bells struck by a leather hammer.

17

Music Styles

"Since music is a language with some meaning at least for the immense majority of mankind, although only a tiny minority of people are capable of formulating a meaning in it, and since it is the only language with the contradictory attributes of being at once intelligible and untranslatable, the musical creator is a being comparable to the gods, and music itself the supreme mystery of the science of man, a mystery that all the various disciplines come up against and which holds the key to their progress."

Claude Levi-Strauss

17.1 Division of the Styles

This chapter presents some basic styles of music. Style is a word that represents the manner by which composers, from different periods and countries present the basic elements of music in their compositions.

In a sense, the music styles are divided into modal, tonal, and serial. The modal music includes African, Indian, Japanese, Arab, Indonesian, and American native traditions, among other cultures. It also includes the ancient Greek musical tradition, and the Gregorian chant. Figure 17.1 illustrates the relation between the basic music parameters, for modal and tonal music.

Tonal music spans the historical period from the medieval polyphony to the atonalism, it covers the Baroque, the Classical, and the Romantic eras. An important aspect of tonal music's syntax is the simultaneous combination of scale tones into chords, to create harmony [Patel, 2010].

Serial music comprises the radical formats of avant-garde music, up to the electronic and the minimalist, or repetitive, music. Serialism is a method of composition that used series of pitches, rhythms, dynamics, timbres, or other musical elements to obtain certain effects [Wisnik, 2017].

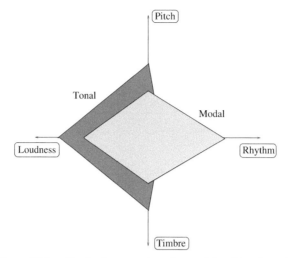

Figure 17.1   The basic parameters, for modal and tonal music.

The basic components of music are [Bennet, 1986]:

Melody – An organized sequence of notes, of different sounds, that makes musical sense to the listener. It is the most important component in a musical piece.

Harmony – The simultaneous sounding of two or more tones, to produce a chord. The chord can be consonant, when the notes sound pleasant or agreeable, or dissonant, when the notes cause tension and a desire to be resolvable to consonant intervals.

Rhythm – The combinations of long and short duration sounds, that convey a sense of movement and regularity in time.

Timbre – The tone color or quality of the sound heard, as a result of the combination of the produced harmonics.

Form – The organization and structure of a composition, and the correlation of musical events within the whole structure.

Tempo – The pace at which music moves according to the speed of the underlying beat. It determines how fast the music is played, it is measured in BPM, and 120 bpm is a common tempo [Buck, 2014].

Dynamics – The varying degrees of volume in the performance of music.

Texture – The way the tempo, melodic, and harmonic materials are combined in a composition to determine the overall quality of the sound in a piece. The texture can be [Yudkin, 2013]:

- Monophonic, when it has a single melodic line, with no harmony at all.
- Polyphonic, or counterpoint, when it has two or more melodic lines.
- Homophonic, when a single melody is heard against a chord accompaniment.
- Song texture, when the melody is accompanied by chords.
- Round, when the musical lines are identical but begin at different times.

17.2 Music Styles

Music is considered an artistic form of acoustic communication that incorporates instrumental or vocal tones in a structured and organized manner. Regarding the stylistic features, some music characteristics are considered. A *genre* is a category of musical composition, or the specific classification of a musical work [University, 2019].

The melody, or harmony, that is based on one of the seven-tone major or minor Western scales is called *diatonic*. The chromatic melodies include notes outside the key of music, and atonal melodies are not based on any key or tonal center.

The combination of intervals in a melody gives it different configurations. For instance, a melody that is not smooth in contour, or has many leaps, is called disjunct. On the other hand, a regular or continuous melody is considered conjunct.

The *texture* is the element that focus on the number of simultaneous musical lines being sounded. A *homophonic texture* is a texture in which voices on different pitches sing the same words simultaneously. It is also the principal melody supported by chord.

17.2.1 Classic Music Styles

Concerto is a general term for a multi-movement work for soloist and orchestra, and concerto grosso is a three-movement work for a small group of soloists and orchestra. The conductor is the leader of a performing group of musicians.

A *symphony* is a multi-movement work for orchestra. A *program symphony* is a programmatic multi-movement work for orchestra, while a *symphonic poem* is a single-movement programmatic work for orchestra.

The *Mass Ordinary*, or simply *Mass*, is a composition based on the five daily prayers of the Roman Catholic mass ordinary – *Kyrie, Gloris, Credo, Sanctus*, and *Agnus Dei*. The Mass traditionally begins with the Greek words *kyrie eleison* for "Lord, have mercy." The *Mass Proper* represents the approximately two dozen prayers of a *Mass*, that change each day to reflect the particular feast day of the liturgical calendar.

An *opera* is a large-scale, fully staged dramatic theatrical work involving solo singers, chorus, and orchestra. It is a form of theatrical piece, descendent from the Greek tragedy, in which music has a leading role and the parts are performed by singers [Sallet, 1976].

An *intermezzo*, also called interlude, is a piece designed originally to be performed between the acts of a play or opera. An *opera buffa* is a comic Italian opera, usually in two acts, on the other hand, an *opera seria* is a serious Italian opera, usually in three acts.

A pastorale is a piece written to imitate the music of shepherds. It is a tender melody usually written in moderate 6/8 or 12/8 time [Wharran, 1969]. *Oratorio* is a large-scale sacred work for solo singers, chorus, and orchestra that is not staged.

Chamber music is a style of music performed by a small group of players, with one player per part. Flamenco is a traditional Spanish folk music, used to accompany singers and dancers [Chapman, 2003].

A marching band is a large ensemble of woodwinds, brass, and percussion used for entertainment at sporting events and parades, usually performing march-like music in a strong duple meter. A concert band, on the other hand, is a large, non-marching, ensemble of woodwind, brass and percussion instruments.

A countermelody is a secondary melodic idea that accompanies and opposes a main thematic idea. The counterpoint is a complex polyphonic texture combining two or more independent melodies.

The expressionism is an ultra-shocking, highly-dissonant modern style of music, based on a similar painting style, from the beginning of the 20th century. It is considered the first European *avant-garde* musical style to focus on subjective styles, that value the emotional expression.

A *fugue* is a complex contrapuntal manipulation of a musical subject. *Incidental music* is a music genre performed during a theatrical play.

Figure 17.2 Portrait of Wilhelm Richard Wagner. (Photographic reproduction of a public domain work of art. Wikimedia Commons.)

Leitmotif is a short musical signature tune associated with a person or concept in a Wagnerian *Musikdrama*, named after the German composer, theatre director, and conductor Wilhelm Richard Wagner (1813–1883). Figure 17.2 shows a portrait of Richard Wagner.

Nocturne is a French term for night piece, a type of character piece for solo piano that evokes the moods and images of night time, usually played as an ensemble piece in several movements during an evening party.

Non-metrical is a type of music without the constraint of a regular beat or steady meter, such as a metronome. It is opposed to metrical music, in which the accents occur at regular intervals and the ratio of the intervals is an integer.

Composers also use metric complex music, in which the ratio of the intervals is integer, and the accents occur at irregular intervals, and jittered non-metric music, in which the accents occur at irregular intervals, but the intervals are randomly jittered.

Polyphony is a music with two or more sounds happening simultaneously. It is opposed to a musical texture with just one voice, known as monophony, or a texture with one dominant melodic voice accompanied by chords, which is called homophony. Therefore, a monophonic texture is single-line texture with no harmony.

A polyphonic texture implies that two or more independent melodic lines sound at the same time, also called polyrhythm. Imitation is a polyphonic texture in which material is presented and then echoed from voice to voice. The *canon* is a type of strict imitation created by strict echoing between a melodic leader and subsequent followers.

A polytonality occurs when music is played in two or more contrasting keys at the same time. The term bitonality is sometimes used, if only two keys are employed in the composition.

A *prelude* is a free-form introductory movement to a fugue or a more complex composition. The term is used in place of overture, introduced by Wagner, Bizet, and other later Romantic composers, to show dramatic unity between the introductory orchestral music and the theatrical drama that follows it.

The *Requiem Mass* is a Roman Catholic mass for the dead, *Missa pro Defunctis* in Latin, that used to be celebrated in the context of a funeral. Originally, such compositions were performed in liturgical service, with monophonic chant and Gregorian melodies.

The *rondo form* is a Classic form in which a main melodic idea returns two or three times in alternation with other melodies, such as ABACA or ABACABA. A musical form comprised of two distinctly opposing sections, "A" versus "B" is said to be in binary form.

Scherzo is a short composition, or a movement from a larger work such as a symphony. It is a rounded binary form, which developed from the minuet, and gradually replaces it as the third, or second, movement in symphonies, string quartets, and sonatas. The word *scherzo* means "joke" in Italian.

A string quartet is a chamber ensemble of two violins, viola and cello, devised in the early Classic era. Also a multi-movement work (genre) for two violins, viola and cello.

The *scerzo and trio form* is the musical movement based on a country dance in triple meter. It was replaced by the aristocratic *minuet* in the early 1800s, as the usual third movement of the Classic four-movement design.

Serenade is a Classic instrumental chamber work similar to a small-scale symphony. It used to be performed for social entertainment of the upper classes. A *sequence* is the immediate repetition of a melodic passage on a higher or lower pitch level.

Solo concerto is a three-movement work for a single soloist *versus* an orchestra.

The *sonata* is considered the highest form of erudite music. It has three or four parts, called movements, that usually obey the following order: *allegro*, *andante*, *minueto*, and *rondo* [de Mattos Priolli, 2013b]. It is also a classic multi-movement work for a piano, or for one instrument with piano accompaniment. The development is the central dramatic section of a sonata form, that moves harmonically through many keys. Also, the process of expanding or manipulation a musical idea.

Sonata form is the common first-movement form of Classic multi-movement instrumental works. Essentially a musical debate between two opposing key centers characterized by three dramatic structural divisions within a single movement: exposition, in which two opposing keys are presented; development, harmonically restless; and recapitulation, all material is presented in the home key. It is also called sonata-allegro form.

The Sonata-rondo form is a formal design that combines aspects of sonata form and rondo form: an ABACABA design in which the opening ABA is the exposition, with two opposing keys presented in "A" *versus* "BA"; C is the development, harmonically restless; and the last ABA is the recapitulation, with all material presented in the home key.

17.2.2 Modern Music Styles

Impressionism is a modern French musical style based on blurred effects, beautiful tone colors and fluid rhythms. It was promoted by Achille-Claude Debussy (1862–1918), a French composer, around the turn of the 1900s.

Atonality is a modern harmony that intentionally avoids a tonal center, or has no apparent home key. This concept was used in the definition of the docecaphony, from the Greek terminology for "twelve sounds." It is also known as twelve-tone technique, twelve-tone serialism or twelve-note composition, and is a method of musical composition first devised by the

Austrian composer Josef Matthias Hauer (1883–1959), who published a manifesto called "law of the twelve tones," in 1919 [Hauer, 2007].

In 1923, Arnold Schönberg (1874–1951) developed a version of the twelve-tone technique. The technique uses all 12 notes of the chromatic scale in the composition, to prevent the emphasis of a specific note. Therefore, the music avoids being in a key. The bitonality is a type of modern music sounding in two different keys simultaneously.

The chromaticism is a harmonic or melodic movement by half-step intervals, and also the harmony that uses pitches beyond the central key of a work [Stokes, 2019]. *Serialism* is a method of modern composition in which the 12 chromatic pitches are put into a numerically ordered series used to control various aspects of a work, involving melody, harmony, tone color, dynamics, instrumentation, etc.

A *program music* is an instrumental music intended to tell a specific story, or set a specific mood or extra-musical image.

The neo-classicism was an early 20th Century compositional style, in which Classic forms and the aesthetics of balance, clarity, and structural unity are combined with modern approaches to harmony, rhythm, and tone color.

17.2.3 Popular Music Styles

Bossa nova was developed by the Brazilian guitarist, singer, and composer João Gilberto Prado Pereira de Oliveira (1931–2019), and popularized in the 1950s and 1960s. It is today one of most celebrated Brazilian music styles. João Gilberto wrote what is considered the first *bossa nova* song, "*Bim Bom*," and his first single release "*Chega De Saudade*" was a hit, as was the long-playing (LP) by the same name released in 1959.

The *bossa nova* rhythm is based on the *samba*, which combines the rhythmic patterns and feel originating in former African slave communities of Bahia, a northeastern Brazilian state. The term bossa nova means literally "new trend" or "new wave" [Castro, 2000].

Stan Getz (1927–1991), an American jazz saxophonist, was born Stanley Gayetski. In the 1960s, he popularized *bossa nova* in North America with the song "The Girl from Ipanema," influenced by João Gilberto and by the Brazilian *maestro*, songwriter, arranger, pianist, and composer Antônio Carlos (Tom) Jobim (1927–1994).

Figure 17.3 shows João Gilberto, playing the acoustic guitar, Antônio Carlos Jobim, at the piano, and Stan Getz, playing the saxophone.

Figure 17.3 João Gilberto, Antônio Carlos Jobim, and Stan Getz. (Adapted from www.wtju.net/jazz-100-hour-59-jazz-bossa-nova/.)

The song "The Girl from Ipanema," (*Garota de Ipanema*, in Portuguese) was composed by Tom Jobim and Vinicius de Moraes (1913–1980), a Brazilian diplomat, poet, and composer, to honor Heloísa (Helô) Eneida Paes Pinto, who lived near the Ipanema beach, in Rio de Janeiro, and was 17 years old at the time. Figure 17.4 shows Vinicius de Moraes and Helô Pinheiro.

Tom Jobim actually fell in love and proposed to her, but she was engaged at the time to the engineer Fernando Mendes Pinheiro, whom she married. She adopted her husband's name, and became Helô Pinheiro. The song was presented for the first time during the musical *O Encontro* (The Date), at the nightclub *Au Bon Gourmet*, in Copacabana, Rio de Janeiro.

Tom Jobim was one of the greatest Brazilian musicians, and made *bossa nova* and international hit with his masterful performance with Frank Sinatra (1915–1998), born Francis Albert Sinatra. Because of the many interpretations of the Brazilian music style by jazz players, and due to certain peculiarities of the *bossa nova* composers and performers, the style became known as an aged blend of *samba* and jazz, with a touch of smoke.

Jazz is a style of American modern popular music combining African and Western musical traits, while hot jazz is the Dixieland style of jazz with a fast *tempo* promoted by the American trumpeter, composer, and vocalist Louis Daniel Armstrong (1901–1971).

Figure 17.4   Vinicius de Moraes and Helô Pinheiro. (Adapted from https://www.culturagenial.com/musica-garota-de-ipanema/.)

A jazz band is an instrumental ensemble comprised of woodwinds, that is, saxophones and clarinets; brasses, that is, trumpets and trombones; and rhythm section, that is piano or guitar, bass, and drum set. *Scat singing* is a style of improvised jazz singing sung on colorful nonsense syllables. A *swing* is a term to describe Big Band jazz music of the 1930s and 1950s. A blending of jazz and rock styles is called fusion.

Rhythm and blues is a style of Afro-American popular music that flourished in the 1940s and 1960s. It is a direct predecessor to rock and roll, a style of popular music that emerged in the 1950s, as a result of the combination of Country-Western, Afro-American, and pop-music elements.

Frevo is a popular music style in Brazil, which is played mainly during carnival time. It is recognized as an intangible cultural heritage of UNESCO, and also as a Brazilian intangible cultural heritage. A curiosity about this style is that its name originated, in the end of the 19th century, from the word fever, because the associated dance style is very energetic, or very hot. Indeed, the verb *frevar*, to dance the *frevo*, is a corruption of the word ferver, that means to boil.

There are three main *frevo* genres, the *frevo de bloco*, or block *frevo*, associated with a big block of dancers; the *frevo de rua*, or street *frevo*, associated with a very fast style of dance [Mendes, 2017]; and the frevo-canção, a somehow slower dance style [Mendes, 2019]. The rhythm is typical from the State of Pernambuco, in the northeast region of Brazil.

The new age was a style of music, popular in the 1980s and 1990s, that rejected the hard-edged beat of rock music by focusing on nature sounds, sweet synthesized tone colors, acoustic instruments, and short hypnotically and repetitive ideas.

Bebop is a complex, highly-improvised style of jazz promoted by Charlie Parker in the 1940s and 1950s, also called Big Band jazz.

Music in which the composition or performance is controlled by a computer is appropriately called computer music.

Ragtime is a style of piano music developed around the turn of the 20th century, with a march-like tempo a syncopated right-hand melody, and a swing bass (oom-pah) left-hand accompaniment.

Rap is a style of popular music developed by Afro-Americans in the 1970s, in which the lyrics are spoken over rhythm tracks. Also called hip hop. Funk is a Brazilian style of popular music developed by Afro-Americans, based on the rap, in which the lyrics are spoken over rhythm tracks.

Yoruba music is a genre of the Yoruba people of Nigeria, Benin and Togo. Known for its strong drumming tradition, the main characteristics of Yoruba music are the prevalence of polyrhythmic drumming and, repetitive call and response structures, which is deeply rooted in African-American musical traditions, that include blues, jazz, gospel, R&B, soul, and hip hop. It is also influential in Latin American and Caribbean musical styles.

17.3 Human Voice Styles

The voice is the human original instrument, and music probably began with the imitation of nature sounds, such as, birds singing. Several styles have been developed through the ages, that represent different forms of singing.

An *aria* is a beautiful manner of solo singing, that includes the use of melismas, repetition, and sequences, and is accompanied by orchestra, with a steady metrical beat. The name derives from the Greek term for air, and is usually part of an opera.

Arias can also be found alone, or in a *cantata*, that is a composition in several movements, written for chorus, soloist, and orchestra. They are, traditionally, religious works. It differs from the *chant*, which is a monophonic melody sung in a free rhythm, such as Gregorian chant of the Roman Catholic church.

The *blues* represents a melancholy style of Afro-American secular music, based on a simple musical and poetic form. The Delta blues style began in the early 1900s, in the Mississippi region. The Classic blues began in the late 1920s, and the Rhythm and Blues in the 1940s.

A *chorale* is a Lutheran liturgical melody, and also a four-part hymn-like chorale harmonization, while a *chorus* is a fairly large choral group, and also a single statement of the main harmonic or melody pattern, such as, in jazz music style.

The *falsetto* is a vocal technique that allows a male to sing in a much higher, lighter register by vibrating only half of the vocal cord. It refers to a type of vocal phonation that allows the singer to reach notes beyond the normal vocal range or *tessitura*.

The *Lied* is the German-texted art song, usually for one voice with piano accompaniment, and describes a form of placing poetry to classical music to create a piece of polyphonic music.

The *madrigal* is a composition genre on a short secular poem, sung by a small group of unaccompanied singers, such as one on a part. The madrigal flourished in Italy from 1520 to 1610, and was adopted in England during the Elizabethan Age (c. 1600) [Wharran, 1969].

The *Gregorian chant* is a monophonic genre, non-metered melodies set to Latin sacred texts. Rhapsody is a free fantasy, usually of heroic or national character, typically brilliant in style. Fantasia is a movement free in spirit and form, such as, an improvisation. Bolero is a rapid Spanish dance in 3/4 time, with accompaniment of castanets [Wharran, 1969].

17.3.1 Voice as an Instrument

The human voice is a sort of reed tone instrument, that is heard in several styles of music. The usual manner to enhance the voice, or tune the instrument, is the study and practice of solfege. Solfege means to sing the written notes, or melodies. It is a fundamental technique to learn how to read and practice music [Med, 1996a].

Tessitura, an Italian term for texture, is the adequate vocal range, or extension, for a singer or musical instrument. It is the range in which the voice presents its characteristic timbre. It is the ambitus in which a particular vocal, or instrumental, part lies. For instance, the guitar *tessitura* comprises more than three and a half-octaves, from the open sixth string, note Mi (E2, corresponding to approximately 82 Hz), to the 18th fretboard, note Si (B5, corresponding to approximately 988 Hz) [Chediak, 1986].

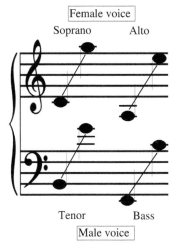

Figure 17.5 Ranges for human voices.

Texture, on the other hand, is how the concepts of *tempo*, melody, and harmony are combined in a composition, to determine the overall quality of the sound in a musical piece.

Figure 17.5 show the approximate ranges of the human voice. As can be expected, the ranges vary according to the bibliographic source.

The *soprano* is the highest ranged woman's voice, or a high pre-pubescent boy's voice. Also the highest-sounding instrument of an instrumental family. Some sopranos are comfortable between C4, the middle C, and A5, the second note *La* above middle C, but it can vary from one singer to another.

The mezzo-soprano is a dramatic woman's voice that combines the power of an alto with the primary high range of a soprano.

Recall that the middle C, the note *Do* in the conventional musical notation, and the fourth C key from left on a standard 88-key piano keyboard, is designated C4 in scientific pitch notation, and corresponds to the frequency 261.53 Hz, if the A440 pitch standard is used.

Alto is a low-ranged female voice, and also the second lowest instrumental range. Altos are comfortable between A3 and C4 [Moulton, 2019].

The *tenor* is a high-ranged male voice. Tenors are comfortable between C3 and F4, the fourth note *Fa* above C4.

The *baritone* is a moderately low male voice, in range between a tenor and a bass. A typical baritone range is from A2, the second A below middle C, to A4, the note *La* above middle C.

The bass is the lowest male voice. It is also the lowest-sounding instrument of an instrumental family. According to *The Oxford Dictionary of Music*, a bass is typically classified as having a range extending from E2, below middle C, to the E4, the note *Mi* above middle C [Kennedy et al., 2013].

Appendix A: Basic Differentiation and Integration Rules

"Music is the arithmetic of sounds as optics is the geometry of light."

Claude Debussy

A.1 Differentiation Rules

The mathematicians Isaac Newton, an English mathematician, physicist, astronomer, and theologian, and Gottfried Wilhelm Leibniz (1646–1716), a German polymath, logician, mathematician, and natural philosopher, disputed for many years over who had first invented calculus.

The derivative of a function measures the sensitivity of the function with respect to a change in its argument. The derivative is a limit of difference quotients of real numbers, in which the numerator is a function of the denominator. The slope of the function $f(x)$ is defined as

$$\frac{\Delta f(x)}{\Delta x} = \frac{f(x + \Delta x) - f(x)}{\Delta x}. \tag{A.1}$$

The limit of the secant line, as the line rotates to touch the curve on a single point, is the tangent line. Therefore, the limit of the difference quotient as Δx approaches zero, if it exists, represents the slope of the tangent line to $(x, f(x))$.

This limit is defined as the derivative of the function $f(\cdot)$ at x,

$$f'(x) = \lim_{x \to 0} \frac{f(x + \Delta x) - f(x)}{\Delta x}. \tag{A.2}$$

Properties of the Derivative

The derivative is a linear operation, if $h(x) = \alpha f(x) + \beta g(x)$, then

$$h'(x) = \alpha f'(x) + \beta g'(x). \tag{A.3}$$

219

If $y = f(x)$ and $x = g(t)$, then the chain rule applies

$$\frac{dy}{dt} = \frac{dy}{dx} \cdot \frac{dx}{dt}. \tag{A.4}$$

The derivative of the multiplication gives

$$\frac{d(y \cdot x)}{dt} = x\frac{dy}{dt} + y\frac{dx}{dt}, \tag{A.5}$$

or, simply

$$d(y \cdot x) = xdy + ydx. \tag{A.6}$$

Derivative of a power

Consider a function,

$$f(x) = x^\alpha, \tag{A.7}$$

in which α is any real number, then

$$f'(x) = \alpha x^{\alpha-1}. \tag{A.8}$$

Derivative of exponential and logarithmic functions

$$\frac{d}{dx}e^{kx} = ke^{kx}, \tag{A.9}$$

in which k is a constant.

$$\frac{d}{dx}\ln(x) = \frac{1}{x}, \qquad x > 0. \tag{A.10}$$

$$\frac{d}{dx}\log_a(x) = \frac{1}{x\ln(a)}. \tag{A.11}$$

Derivative of trigonometric functions

$$\frac{d}{dx}\sin(kx) = k\cos(kx). \tag{A.12}$$

$$\frac{d}{dx}\cos(kx) = -k\sin(kx). \tag{A.13}$$

Therefore,

$$\frac{d^2}{dx^2}\sin(kx) = -k^2\sin(kx). \tag{A.14}$$

$$\frac{d}{dx}\tan(x) = \sec^2(x) = \frac{1}{\cos^2(x)} = [1+\tan^2(x)]. \tag{A.15}$$

Derivative of the inverse trigonometric functions

$$\frac{d}{dx}\arcsin(x) = \frac{1}{\sqrt{1-x^2}}, \qquad -1 < x < 1. \tag{A.16}$$

$$\frac{d}{dx}\arccos(x) = -\frac{1}{\sqrt{1-x^2}}, \qquad -1 < x < 1. \tag{A.17}$$

$$\frac{d}{dx}\arctan(x) = \frac{1}{1+x^2}. \tag{A.18}$$

A.2 Integration Rules

The first technique to compute integrals was the method of exhaustion, developed by the Greek astronomer Eudoxus of Cnidus (c.390 BC–c.337 BC), an astronomer, mathematician, and scholar, to find the area of the circle by dividing it into an infinite number of triangles for which the area was known, since Pythagoras.

This method was improved by Archimedes, in the 3rd century BC, and used to calculate the area of the exponential and parabolic curves. The modern notation for the indefinite integral was introduced by Gottfried Wilhelm Leibniz.

The integral is the inverse function of the derivative. It is defined as,

$$F(x) = \int f(x)dx, \tag{A.19}$$

for a function $f(x)$ subject to some constraints. Because the derivative of the constant is zero, it is necessary to add a constant C to the indefinite integral, as follows.

Integral of a constant

$$\int kdx = kx + C. \tag{A.20}$$

Integral of the identity function

$$\int x dx = \frac{x^2}{2} + C. \qquad (A.21)$$

Integral of a polynomial function

$$\int x^n dx = \frac{x^{n+1}}{n+1} + C. \qquad (A.22)$$

Integral of trigonometric functions

$$\int \sin(kx) dx = -\frac{\cos(kx)}{k} + C. \qquad (A.23)$$

$$\int \cos(kx) dx = \frac{\sin(kx)}{k} + C. \qquad (A.24)$$

Integral of the exponential function

$$\int e^{kx} dx = \frac{e^{kx}}{k} + C. \qquad (A.25)$$

The modern notation for the definite integral, with limits above and below the integral sign, was first used by Jean-Baptiste Joseph Fourier.

$$\int_a^b f(x) dx = [F(x)]_a^b = F(b) - F(a). \qquad (A.26)$$

Appendix B: Fourier Series and Transforms

"Music is like a dream. One that I cannot hear."

Ludwig van Beethoven

This appendix presents some properties of Fourier series, and Fourier transform [Hsu, 1973; Spiegel, 1976; Baskakov, 1986; Haykin, 1987; Lathi, 1989; Gradshteyn and Ryzhik, 1990; Oberhettinger, 1990; Alencar and da Rocha Jr., 2005].

Trigonometric series: $f(t) = a_0 + \sum_{n=1}^{\infty} (a_n \cos n\omega_0 t + b_n \sin n\omega_0 t)$

Cosine series: $f(t) = C_0 + \sum_{n=1}^{\infty} C_n \cos(n\omega_0 t + \theta_n)$

Complex series: $f(t) = \sum_{n=-\infty}^{\infty} F_n e^{jn\omega_0 t}$, $f(t+T) = f(t)$, $\omega_0 = \dfrac{2\pi}{T}$.

Conversion formulas:

$$F_0 = a_0, \qquad F_n = \frac{1}{2}(a_n - jb_n), \qquad F_{-n} = \frac{1}{2}(a_n + jb_n),$$

$$F_n = |F_n|e^{j\phi_n}, \qquad |F_n| = \frac{1}{2}\sqrt{a_n^2 + b_n^2},$$

$$a_0 = F_0, \qquad a_n = F_n + F_{-n}, \qquad b_n = j(F_n - F_{-n}),$$

$$C_0 = a_0, \qquad C_n = 2|F_n| = \sqrt{a_n^2 + b_n^2}, \qquad \theta_n = -\arctan\left(\frac{b_n}{a_n}\right).$$

Definitions of Fourier transforms and some properties:

- Definition of Fourier transform: $F(\omega) = \int_{-\infty}^{\infty} f(t)e^{-j\omega t}dt$.

- Inverse of Fourier transform: $f(t) = \dfrac{1}{2\pi} \int_{-\infty}^{\infty} F(\omega)e^{j\omega t}d\omega$.

- Linearity of Fourier transform: $\alpha f(t) + \beta g(t) \leftrightarrow \alpha F(\omega) + \beta G(\omega)$.

- Magnitude and phase of the transform: $F(\omega) = |F(\omega)|e^{j\theta(\omega)}$.

- Transform of an even function $f(t)$: $F(\omega) = 2\int_0^{\infty} f(t)\cos\omega t dt$.

- Transform of an odd function $f(t)$: $F(\omega) = -2j\int_0^{\infty} f(t)\sin\omega t dt$.

- Area under function in the time domain: $F(0) = \int_{-\infty}^{\infty} f(t)dt$.

- Area under transform in the frequency domain: $f(0) = \dfrac{1}{2\pi}\int_{-\infty}^{\infty} F(\omega)d\omega$.

Parseval's theorem:

$$\int_{-\infty}^{\infty} f(t)g(t)dt = \frac{1}{2\pi}\int_{-\infty}^{\infty} F(\omega)G^*(\omega)d\omega,$$

$$\int_{-\infty}^{\infty} |f(t)|^2 dt = \frac{1}{2\pi}\int_{-\infty}^{\infty} |F(\omega)|^2 d\omega,$$

$$\int_{-\infty}^{\infty} f(\omega)G(\omega)d\omega = \int_{-\infty}^{\infty} F(\omega)g(\omega)d\omega.$$

Fourier Transforms

$f(t)$	$F(\omega)$
$F(t)$	$2\pi f(-\omega)$
$f(t-\tau)$	$F(\omega)e^{-j\omega\tau}$
$f(t)e^{j\omega_0 t}$	$F(\omega - \omega_0)$
$f(at)$	$\dfrac{1}{|a|}F(\dfrac{\omega}{a})$
$f(-t)$	$F(-\omega)$
$f^*(t)$	$F^*(-\omega)$
$f(t)\cos\omega_0 t$	$\dfrac{1}{2}F(\omega - \omega_0) + \dfrac{1}{2}F(\omega + \omega_0)$
$f(t)\sin\omega_0 t$	$\dfrac{1}{2j}F(\omega - \omega_o) - \dfrac{1}{2j}F(\omega + \omega_0)$
$f'(t)$	$j\omega F(\omega)$
$f^{(n)}(t)$	$(j\omega)^n F(\omega)$
$\displaystyle\int_{-\infty}^{t} f(x)dx$	$\dfrac{1}{j\omega}F(\omega) + \pi F(0)\delta(\omega)$
$-jtf(t)$	$F'(\omega)$
$(-jt)^n f(t)$	$F^{(n)}(\omega)$
$f(t) * g(t) = \displaystyle\int_{-\infty}^{\infty} f(\tau)g(t-\tau)d\tau$	$F(\omega)G(\omega)$

Fourier Transforms

$f(t)$	$F(\omega)$
$\delta(t)$	1
$\delta(t - \tau)$	$e^{-j\omega\tau}$
$\delta'(t)$	$j\omega$
$\delta^{(n)}(t)$	$(j\omega)^n$
$e^{-at}u(t)$	$\dfrac{1}{a + j\omega}$
$e^{-a|t|}$	$\dfrac{2a}{a^2 + \omega^2}$
e^{-at^2}	$\sqrt{\dfrac{\pi}{a}}e^{-\omega^2/(4a)}$
te^{-at^2}	$j\sqrt{\dfrac{\pi}{4a^3}}\omega e^{-\omega^2/(4a)}$
$te^{-at}u(t)$	$\dfrac{1}{(a + j\omega)^2}$
$\dfrac{t^{n-1}}{(n-1)!}e^{-at}u(t)$	$\dfrac{1}{(a + j\omega)^n}$
$p_T(t) = \begin{cases} A & \text{for } |t| \le T/2 \\ 0 & \text{for } |t| > T/2 \end{cases}$	$AT\dfrac{\sin\left(\dfrac{\omega T}{2}\right)}{\left(\dfrac{\omega T}{2}\right)}$
$\dfrac{\sin at}{\pi t}$	$p_{2a}(\omega)$
$e^{-at}\sin bt\, u(t)$	$\dfrac{b}{(a + j\omega)^2 + b^2}$
$e^{-at}\cos bt\, u(t)$	$\dfrac{a + j\omega}{(a + j\omega)^2 + b^2}$

Fourier Transforms

$f(t)$	$F(\omega)$
$\dfrac{1}{a^2 + t^2}$	$\dfrac{\pi}{a} e^{-a|\omega|}$
$\dfrac{t}{a^2 + t^2}$	$j\pi e^{-a|\omega|}[u(-\omega) - u(\omega)]$
$\dfrac{\cos bt}{a^2 + t^2}$	$\dfrac{\pi}{2a}[e^{-a|\omega-b|} + e^{-a|\omega+b|}]$
$\dfrac{\sin bt}{a^2 + t^2}$	$\dfrac{\pi}{2aj}[e^{-a|\omega-b|} - e^{-a|\omega+b|}]$
$\sin bt^2$	$\dfrac{\pi}{2b}\left[\cos\dfrac{\omega^2}{4b} - \sin\dfrac{\omega^2}{4b}\right]$
$\cos bt^2$	$\dfrac{\pi}{2b}\left[\cos\dfrac{\omega^2}{4b} + \sin\dfrac{\omega^2}{4b}\right]$
$\operatorname{sech} bt$	$\dfrac{\pi}{b}\operatorname{sech}\dfrac{\pi\omega}{2b}$
$\ln\left[\dfrac{x^2 + a^2}{x^2 + b^2}\right]$	$\dfrac{2e^{-b\omega} - 2e^{-a\omega}}{\pi\omega}$
$f(t)g(t)$	$\dfrac{1}{2\pi}F(\omega) * G(\omega) = \dfrac{1}{2\pi}\displaystyle\int_{-\infty}^{\infty} F(\phi)G(\omega - \phi)d\phi$
$e^{j\omega_0 t}$	$2\pi\delta(\omega - \omega_0)$
$\cos\omega_0 t$	$\pi[\delta(\omega - \omega_0) + \delta(\omega + \omega_0)]$
$\sin\omega_0 t$	$-j\pi[\delta(\omega - \omega_0) - \delta(\omega + \omega_0)]$
$\sin\omega_0 t u(t)$	$\dfrac{\omega_0}{\omega_0^2 - \omega^2} + \dfrac{\pi}{2j}[\delta(\omega - \omega_0) - \delta(\omega + \omega_0)]$
$\cos\omega_0 t u(t)$	$\dfrac{j\omega}{\omega_0^2 - \omega^2} + \dfrac{\pi}{2}[\delta(\omega - \omega_0) + \delta(\omega + \omega_0)]$

Fourier Transforms

$f(t)$	$F(\omega)$
$u(t)$	$\pi\delta(\omega) + \dfrac{1}{j\omega}$
$u(t-\tau)$	$\pi\delta(\omega) + \dfrac{1}{j\omega}e^{-j\omega\tau}$
$u(t) - u(-t)$	$\dfrac{2}{j\omega}$
$tu(t)$	$j\pi\delta'(\omega) - \dfrac{1}{\omega^2}$
1	$2\pi\delta(\omega)$
t	$2\pi j\delta'(\omega)$
t^n	$2\pi j^n \delta^{(n)}(\omega)$
$|t|$	$\dfrac{-2}{\omega^2}$
$\dfrac{1}{t}$	$\pi j - 2\pi j u(\omega)$
$\dfrac{1}{t^n}$	$\dfrac{(-j\omega)^{n-1}}{(n-1)!}[\pi j - 2\pi j u(\omega)]$
$\dfrac{1}{e^{2t}-1}$	$\dfrac{-j\pi}{2}\coth\dfrac{\pi\omega}{2} + \dfrac{j}{\omega}$
$f_E(t) = \dfrac{1}{2}[f(t) + f(-t)]$	$\mathrm{Re}\,(\omega)$
$f_O(t) = \dfrac{1}{2}[f(t) - f(-t)]$	$j\mathrm{Im}\,(\omega)$
$f(t) = f_E(t) + f_O(t)$	$F(\omega) = \mathrm{Re}\,(\omega) + j\mathrm{Im}(\omega)$

Appendix C: Notes, Frequencies, and Wavelengths

> *"Music is enough for a lifetime, but a lifetime is not enough for music."*
>
> *Sergei Rachmaninoff*

This appendix presents the frequencies and wavelengths for the equal-tempered scale, which is the common musical scale for the tuning of guitars, pianos, and other instruments of fixed scale. The equal temperament divides the octave into 12 semitones, and the frequency interval between every pair of adjacent notes has the same ratio, therefore, the pitch is perceived as the logarithm of frequency [Palmer et al., 1994].

Figure C.1 shows the 12-tone equal temperament chromatic scale on C, one full octave ascending, notated only with sharps.

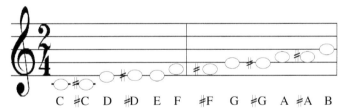

C ♯C D ♯D E F ♯F G ♯G A ♯A B

Figure C.1   The 12-tone equal temperament chromatic scale on C.

Table C.1   Frequencies and wavelengths for the equal-tempered scale, A4 = 440 Hz, and the middle C is C4. (Adapted from Physics of Music-Notes – http://pages.mtu.edu/~suits/notefreqs.html)

Notes	Frequencies (Hz)	Wavelengths (cm)
C0	16.35	2109.89
C#0/Db0	17.32	1991.47
D0	18.35	1879.69
D#0/Eb0	19.45	1774.20
E0	20.60	1674.62
F0	21.83	1580.63
F#0/Gb0	23.12	1491.91
G0	24.50	1408.18
G#0/Ab0	25.96	1329.14
A0	27.50	1254.55
A#0/Bb0	29.14	1184.13
B0	30.87	1117.67
C1	32.70	1054.94
C#1/Db1	34.65	995.73
D1	36.71	939.85
D#1/Eb1	38.89	887.10
E1	41.20	837.31
F1	43.65	790.31
F#1/Gb1	46.25	745.96
G1	49.00	704.09
G#1/Ab1	51.91	664.57
A1	55.00	627.27
A#1/Bb1	58.27	592.07
B1	61.74	558.84
C2	65.41	527.47
C#2/Db2	69.30	497.87
D2	73.42	469.92
D#2/Eb2	77.78	443.55
E2	82.41	418.65

Table C.2 Frequencies and wavelengths for the equal-tempered scale, A4 = 440 Hz, and the middle C is C4. (Adapted from Physics of Music-Notes – http://pages.mtu.edu/~suits/notefreqs.html)

Notes	Frequencies (Hz)	Wavelengths (cm)
F2	87.31	395.16
F#2/Gb2	92.50	372.98
G2	98.00	352.04
G#2/Ab2	103.83	332.29
A2	110.00	313.64
A#2/Bb2	116.54	296.03
B2	123.47	279.42
C3	130.81	263.74
C#3/Db3	138.59	248.93
D3	146.83	234.96
D#3/Eb3	155.56	221.77
E3	164.81	209.33
F3	174.61	197.58
F#3/Gb3	185.00	186.49
G3	196.00	176.02
G#3/Ab3	207.65	166.14
A3	220.00	156.82
A#3/Bb3	233.08	148.02
B3	246.94	139.71
C4	261.63	131.87
C#4/Db4	277.18	124.47
D4	293.66	117.48
D#4/Eb4	311.13	110.89
E4	329.63	104.66
F4	349.23	98.79
F#4/Gb4	369.99	93.24
G4	392.00	88.01
G#4/Ab4	415.30	83.07

Table C.3   Frequencies and wavelengths for the equal-tempered scale, A4 = 440 Hz, and the middle C is C4. (Adapted from Physics of Music-Notes – http://pages.mtu.edu/~suits/notefreqs.html)

Notes	Frequencies (Hz)	Wavelengths (cm)
A4	440.00	78.41
A#4/Bb4	466.16	74.01
B4	493.88	69.85
C5	523.25	65.93
C#5/Db5	554.37	62.23
D5	587.33	58.74
D#5/Eb5	622.25	55.44
E5	659.25	52.33
F5	698.46	49.39
F#5/Gb5	739.99	46.62
G5	783.99	44.01
G#5/Ab5	830.61	41.54
A5	880.00	39.20
A#5/Bb5	932.33	37.00
B5	987.77	34.93
C6	1046.50	32.97
C#6/Db6	1108.73	31.12
D6	1174.66	29.37
D#6/Eb6	1244.51	27.72
E6	1318.51	26.17
F6	1396.91	24.70
F#6/Gb6	1479.98	23.31
G6	1567.98	22.00
G#6/Ab6	1661.22	20.77
A6	1760.00	19.60
A#6/Bb6	1864.66	18.50
B6	1975.53	17.46
C7	2093.00	16.48
C#7/Db7	2217.46	15.56

Table C.4 Frequencies and wavelengths for the equal-tempered scale, A4 = 440 Hz, and the middle C is C4. (Adapted from Physics of Music-Notes – http://pages.mtu.edu/~suits/notefreqs.html)

Notes	Frequencies (Hz)	Wavelengths (cm)
D7	2349.32	14.69
D#7/Eb7	2489.02	13.86
E7	2637.02	13.08
F7	2793.83	12.35
F#7/Gb7	2959.96	11.66
G7	3135.96	11.00
G#7/Ab7	3322.44	10.38
A7	3520.00	9.80
A#7/Bb7	3729.31	9.25
B7	3951.07	8.73
C8	4186.01	8.24
C#8/Db8	4434.92	7.78
D8	4698.63	7.34
D#8/Eb8	4978.03	6.93
E8	5274.04	6.54
F8	5587.65	6.17
F#8/Gb8	5919.91	5.83
G8	6271.93	5.50
G#8/Ab8	6644.88	5.19
A8	7040.00	4.90
A#8/Bb8	7458.62	4.63
B8	7902.13	4.37

Glossary of Musical Terminology

"And those who were seen dancing were thought to be insane by those who could not hear the music."

Friedrich Nietzsche

This glossary presents the main definitions, common acronyms and the terminology used in music theory [de la Fuente, 2014; Davis and Davis, 1987; Wharran, 1969; Capone, 2007; Chapman, 2003; Neely, 1988; Burrows, 2002; Burrows, 2004; Loy, 2011b; Loy, 2011a; Franceschina, 2015; Benade, 1990; Abromont and de Montalem bert, 2010; Miller, 2005; Yudkin, 2013; Ball, 2010; Rodriguez, 2017; Hodge, 2006; Howard, 1991; Bennett, 1998; Károlyi, 2002; Friedland, 2004; Wade-Matthews and Thompson, 2003; de Lacerda, 1966; Pen, 1992; Levitin, 2006; Schonbrun, 2014; Moon, 2015; Martineau, 2008; Almada, 2009; University, 2019; Stokes, 2019; Pilhofer and Day, 2013; de Paula, 2007; Med, 1996b; Patel, 2010].

A440 – The pitch standard established by the International System of Units (SI) for orchestras, corresponding to 440 Hz.

ABA form – Tripartite form, found in opera arias and some slow movement instrumental works. It features and opening A section, a contrasting B section, and a return to the A section.

Absolute music – Instrumental music with no intended story, or non-programmatic music.

Absolute pitch – The ability to accurately identify the pitch of a tone, without resorting to an external reference pitch.

A cappella – Choral music with no instrumental accompaniment.

Accelerando – Gradually accelerating the speed of the rhythmic beat.

Accent – An instantaneous emphasis or stress on a note or chord, using either a dynamic attack (louder volume), or *agogic* attack, greater duration.

Accidental – Any of the five symbols (double flat ♭♭, flat ♭, sharp ♯, double sharp ×, natural ♮) that lower or raise a pitch by one or two semitones, usually used to alter or restore a key.

Adagio – A slow tempo.

Ad libitum – Instruction that the performer may freely interpret or improvise a passage of a music piece.

Aerophone – A musical instrument that produces sound primarily by the vibration of the air, without the use of strings or membranes, and without the vibration of the instrument itself.

Allegro – A fast tempo.

Alto – A low-ranged female voice, also called alto voice. The second lowest instrumental range.

Alto clef – A clef used mainly by the viola, that places middle Do (middle C) on the middle line of the staff.

Amplitude – A distance measure in a wave, from its equilibrium, or zero level, to its peak, or maximum level.

Andante – A moderate tempo. Also implying a walking speed, because *andare* means to walk, in Italian.

Anticipation – Arrive at the target note or chord before the beat, usually a half a beat earlier.

Antinode – A point where displacement produced by vibration is greatest.

ANSI – American National Standards Institute.

Approach note – A note that leads up or down to a structural tone.

Arranging – The same as orchestration, the art of scoring music for an orchestra or band.

Aria – A beautiful manner of solo singing, accompanied by orchestra, with a steady metrical beat. Also, a lyrical section of opera for solo singer and orchestra.

Arpeggio – A chord that is played one note at a time, usually in ascending or descending order.

Art-music – A general term used to describe the formal concert music traditions of the West, as opposed to popular and commercial music styles.

A tempo – It is a return to the original tempo, that usually appears after a *accelerando* ou *ritardando*.

Art song – A musical setting of artistic poetry for solo voice accompanied by piano or orchestra.

Attack – Manner of emitting a sound, which can be *legato*, *staccato*, *portato*, etc. In acoustics, it is the moment the sound reaches resonance, and produces several harmonics.

Attack time – The interval required, after a sudden increase in input signal amplitude, to attain a certain percentage of the signal peak.

Attenuation – A decrease in signal magnitude or power.

Atonality – Modern harmony that intentionally avoids a tonal center, or has no apparent home key.

Audio frequency – Any frequency that can be heard by the human ear. The range of audio frequencies varies from person to person, but it is considered to lie between 20 Hz and 20,000 Hz.

Augmentation – Lengthening the rhythmic values of a fugal subject.

Augmentation dot – Increases the duration of a note by one half.

Augmented chord – One of the four basic chord types, composed of the root, major third, and augmented fifth of the major scale.

Augmented interval – A interval that is one half-step larger than major or perfect.

Avant-garde – A French term that means at the forefront, and describes highly experimental modern musical styles.

Back beat – A term used to describe the emphasis of the weak beats two and four, in common time (4/4).

Ballet – A programmatic theatrical work for dancers and orchestra.

Bandwidth – Distance between the upper and lower frequency limits of a sound.

Bar – A common term for a musical measure. A distinct measurement of beats, which is dictated by the time signature. The end of the bar is indicated by a vertical line that runs through the staff. The most common grouping of beats is four in a bar, called common time.

Baritone – A moderately low male voice. It is in range between a tenor and a bass.

Baroque era – A musical period (c.1600–1750), of extremely ornate and elaborate approaches to the arts. This era saw the rise of instrumental music, the invention of the modern violin family, and the creation of the first orchestras, by Vivaldi, Handel, and Bach.

Basilar membrane – A thin membrane within the cochlea, inside the inner ear, that embeds the hair cells.

Bass – The lowest male voice, or the lowest pitch of a chord.

Bass clef – A symbol put on the staff that indicates the use of instruments that have a lower pitch. It is also called the F clef.

Bass drum – The lowest-sounding non-pitched percussion instrument.

Basso continuo – The back-up ensemble of the Baroque era, usually comprised of a keyboard instrument, a harpsichord or an organ, and a melodic stringed bass instrument, a viol' da gamba or cello.

Bassoon – The lowest-sounding regular instrument of the woodwind family. It is a double-reed instrument.

Beam – A horizontal line that connects and replaces a group of short notes.

Beat – A musical pulse. A fundamental unit of time measurement.

Bebop – A complex, highly-improvised style of jazz promoted by Charlie Parker in the 1940s and 1950s, also called Big Band jazz.

Bemol – A musical symbol that lowers the pitch one half-step. It is represented by the symbol (♭), a stylistic abbreviation of the word *bemol*.

Binary form – A musical form comprised of two distinctly opposing sections, such as, A versus B, which can be repeated, as in the pattern AABB.

Bis – An instruction to repeat a short passage.

Bitonality – Modern music sounding in two different keys simultaneously.

Blue note – Flattened note played or sung in jazz.

Blues – A melancholy style of Afro-American secular music, based on a simple musical and poetic form. The Delta blues style began in the early 1900s, in the Mississipi region. The Classic blues began in the late 1920s, and the Rhythm and Blues in the 1940s.

Blues progression – A 12-bar sequence of chords that is usual in blues and jazz music, as follows: I-I-I-I-IV-IV-I-I-V7-IV-I-I.

Barre – The method of placing the first finger, or other fingers, across adjacent strings to hold down adjacent chord notes.

Bolero – A rapid Spanish dance in 3/4 time, with accompaniment of castanets.

Bout – A term applied to the upper and lower sections of the guitar body.

Brass instrument – A powerful metallic instrument with a mouthpiece and tubing that must be blown into by the player, such as trumpet, trombone, French horn, tuba, baritone, and bugel.

Bridge – A short section that links two important parts of a piece of music.

Cadence – A melodic or harmonic punctuation mark at the end of a phrase, major section or entire work.

Cadenza – An unaccompanied section of virtuosistic display played by a soloist in a concerto.

Call and response – A traditional African process in which a leader's phrase, dubbed a call, is repeatedly answered by a chorus. This process became an important aspect of many Afro-American styles.

Camber – Curvature of the fingerboard, also known as the radius.

Canon – A type of strict imitation created by strict echoing between a melodic leader and subsequent followers.

Cantata – A composition in several movements, written for chorus, soloist, and orchestra. They are, traditionally, religious works.

Cello – The tenor-ranged instrument of the modern string family. It is an abbreviation for violoncello.

Chamber music – Music performed by a small group of players, with one player per part.

Chamber orchestra – A smaller version of the complete orchestra.

Chance music – A modern manner of composition in which some or all of the work is left to chance.

Chant – A monophonic melody sung in a free rhythm, such as Gregorian chant of the Roman Catholic church.

Character piece – A one-movement programmatic work for a solo pianist.

Chimes – A percussion instrument comprised of several tube-shaped bells struck by a leather hammer.

Chorale – A Lutheran liturgical melody, and also a four-part hymn-like chorale harmonization.

Chord – A harmonic combination that has three or more pitches sounding simultaneously. Also, three or more notes played together at the same time. The basic chord is the triad.

Chordophone – A musical instrument that relies on the vibration of a set of strings, such as, a violin, or a cello.

Chord progression – A sequence of chords played in a song or in a musical phrase.

Chorus – A fairly large choral group, and also a single statement of the main harmonic or melody pattern, in Jazz.

Chromaticism – Harmonic or melodic movement by half-step intervals, and also the harmony that uses pitches beyond the central key of a work.

Chromatic note – A note that is taken from outside of a given major scale.

Chromatic scale – A scale composed of 12 tones, each separated by a semitone.

Circle of fifths – A pattern that is used to study the relationship between the keys. It is a sequence that begins with a tonic, followed by a dominant, and so on.

Clarinet – The tenor-ranged instrument of the woodwind family. It has a single reed instrument.

Classic era – A politically turbulent era i(c.1750–1820), focused on structural unity, clarity, and balance. Great composers, such as Haydn, Mozart, and Beethoven belong to that era.

Clef – From the French word *clé*, for key, is a musical symbol used to indicate the pitch of written notes.

Coda – A concluding section appended to the end of a work. It means tail in Italian.

Collegium musicum – A university ensemble dedicated to the performance of early music (before 1750).

Color – Refers to the variety of timbres used.

Compass – Rhythmic organization of the music, in groups of two or more beats, or pulses.

Compound interval – An interval larger than an octave.

Composition – Musical composition refers to an original piece of music, either vocal or instrumental. It is also the structure of a musical piece, or the process of creating or writing a new piece of music.

Computer music – Music in which the composition or performance is controlled by a computer.

Concert band – A large, non-marching, ensemble of woodwind, brass, and percussion instruments.

Concerto – The general term for a multi-movement work for soloist and orchestra.

Concerto grosso – A three-movement work for a small group of soloists and orchestra.

Conductor – The leader of a performing group of musicians.

Consonance – Pleasant-sounding harmony. Pythagoras established the first rules, regarding the sections of a string, to obtain consonance.

Contrabassoon – The lowest-sounding double-reed instrument of the wood-wind family cool jazz. Also a relaxed style of modern jazz, promoted in the 1950s and 1960s by Brubeck and others.

Contour – The characteristic shape of a melodic line.

Cornet – A mellow-sounding member of the trumpet family.

Countermelody – A secondary melodic idea that accompanies and opposes a main thematic idea.

Counterpoint – A complex polyphonic texture combining two or more independent melodies.

Crotchet – A note worth one beat within a bar of four-four tone.

Crescendo – The sound intensity gradually getting louder.

Cycle of fifths – It is a didactic system to represent the 12 perfect fifths of the equal tempered chromatic scale.

Cymbals – Percussion instrument usually consisting of two circular brass plates struck together as a pair.

Da capo – It is an Italian expression that means from the head, that is, a written indication telling a performer to go back to the start of a piece.

Da capo al fine – It is an Italian phrase meaning from the beginning to the end of a piece.

Dal segno – Means from the sign, indicating that the performer must repeat a sequence from a point marked by a symbol.

Decibel – A logarithmic scale used to measure the sound intensity.

Decrescendo – The sound intensity is gradually getting quieter.

Development – The central dramatic section of a sonata form that moves harmonically through many keys. Also, the process of expanding, or manipulation a musical idea.

Diatonic – A melody or harmony based on one of the seven-tone major or minor Western scales. Also called natural scale.

Diabolus in Musica – The name given to the tritone, an interval of three whole steps, in the 17th century. Because it was considered dissonant, the use of the interval was prohibited by music theorists of the time.

Dies Irae – A chant from the Requiem mass dealing with God's wrath on the day of judgment.

Diminished chord – One of the four basic chord types, composed of the root, minor third, and diminished fifth degrees of the major scale.

Diminuendo – Gradually getting quieter (cf. Decrescendo).

Diminution – To shorten the note values of a theme, usually to render it twice as fast.

Disjunct – A melody that is not smooth in contour, or has many leaps.

Dissonance – A combination of tones that sounds discordant and unstable, that needs resolution to a more stable and pleasing harmony.

Distortion – An undesired change in the signal waveform, usually caused by a system nonlinearity.

Doctrine of affections – The Baroque methodology for evoking a specific emotion through music and text.

Dominant – The fifth chord, degree, or tone of a scale.

Dotted note – A written note with a dot to the right of it. The dot adds half the rhythmic duration to the note's original value.

Double bass – The lowest-sounding instrument of the modern string family.

Downbeat – The first beat of a musical measure, usually accented more strongly than other beats.

Duet – A musical composition of two performers.

Duple meter – A basic metrical pattern having two beats per measure.

Dynamics – The musical element of relative musical loudness, such as *forte*, or quietness, such as *piano*.

Echo – A reflection of a wave, which returns with sufficient magnitude and delay to be perceived by the human hearing.

Electric instrument – An instrument whose sound is produced or modified by an electro-magnetic pick up.

Electronic instrument – An instrument whose sound is produced, or modified by, electronic means.

Enharmonic – Two notes that sound the same, but are different when written. For example, Do♯ (C♯) and Re♭ (D♭) are enharmonic notes.

Electrophone – A musical instrument in which the original sound either is produced by electronic means, or is generated in a conventional manner and electronically amplified. The electronic music synthesizer and the electronic organ are examples.

English horn – A tenor oboe. Also, a richly nasal-sounding double-reed woodwind instrument.

Ensemble – A group of musical performers.

Envelope – The characteristic manner that the sound intensity of a note changes with time.

Episode – An intermediary, contrasting, section of a Baroque fugue or Classic rondo form equal temperament. Also, the standard modern tuning system in which the octave is divided into 12 equal half-steps.

Equalizer – A device designed to compensate for an undesired signal characteristic, usually in amplitude, frequency, or phase.

Equal temperament – The equal temperament divides the octave into 12 semitones, and the frequency interval between every pair of adjacent notes has the same ratio, therefore, the pitch is perceived as the logarithm of frequency.

Étude – A French word for a study piece, designed to help a performer master a particular technique.

Exposition – The opening section of a fugue. Also the opening section of a Classic sonata form, in which the two opposing key centers are exposed to the listener for the first time.

Expression mark – Words or symbols written on a score to guide the player on dynamics, articulation, and tempo.

Expressionism – An ultra-shocking, highly-dissonant modern style of music, based on a similar painting style.

Falsetto – A vocal technique that allows a male to sing in a much higher, lighter register by vibrating only half of the vocal cord.

Fanfare – A stirring phrase, or a complete short work, that is written originally, or entirely for brass instruments.

Fantasia – A movement free in spirit and form, such as, an improvisation.

Flamenco – A traditional Spanish folk music, used to accompany singers and dancers.

Flat sign – A musical symbol that lowers the pitch one half-step. It is represented by the symbol (♭), a stylistic abbreviation of the word *bemol*.

Flute – A metal tubular instrument that is the soprano instrument of the standard woodwind family.

Figure – Every form of writing a note pitch, duration, or silence.

Film music – A music genre that serves either as background or foreground material for a movie.

Filter – A circuit, or equipment, that separates waves on the basis of their frequencies.

Form – The elemental category describing the shape or design of a musical work or movement.

Formant – A group of frequencies of a certain bandwidth that is emphasized by a resonant system.

Forte – A loud dynamic marking (f).

Fortepiano – An early prototype of the modern piano, designed to play both loud and quiet.

Fortissimo – A very loud dynamic marking (ff).

French horn – A valved brass instrument of medium to medium-low range, that is alto to bass.

Fret – A small metal strip inserted into the fingerboard of some instruments, such as guitars, to make pitch definition more precise.

Fugue – A style of contrapuntal composition, or manipulation, of a musical subject.

Fusion – A blending of jazz and rock styles.

Gain – The increase in signal power caused by an amplifier.

Gamelan – An Indonesian musical ensemble comprised primarily of percussion instruments.

Gamut – The whole range of frequencies reachable by an instrument or voice.

Genre – A category of musical composition, or the specific classification of a musical work.

Glissando – A rapid slide between two distant pitches.

Glockenspiel – A pitched-percussion instrument comprised of metal bars in a frame struck by a mallet.

Gong – A non-pitched percussion instrument made of a large metal plate struck with a mallet

Grave – A slow, solemn tempo.

Gregorian chant – Monophonic genre, non-metered melodies set to Latin sacred texts. Also called plainchant.

Guitar – A six-stringed fretted instrument.

Habañera – An exotic Cuban dance in duple meter.

Half step – The smallest interval in the Western system of equal temperament.

Harmonics – The integer multiples of the fundamental tone, also called overtones.

Harmony – The elemental category describing vertical combinations of pitches. Also, the accompanying parts behind the main melody.

Harmonization – The choice of chords to accompany a melodic line.

Harp – A plucked instrument having strings stretched on a triangular frame.

Harpsichord – An ancient keyboard instrument, whose sound is produced by a system of levered picks that pluck its metal strings. It was common in the Renaissance and Baroque eras.

Hearing limit – The intensity above which the sound is perceived as pain, possibly damaging to the hearing system.

Holistic composition – The act of creating the chords and the melody of a musical composition simultaneously.

Home key – The same as tonic key.

Homophonic texture – A main melody supported by chord. Also, a texture in which voices on different pitches sing the same words simultaneously.

Horn – Short for French horn.

Hot jazz – A Dixieland style of jazz with a fast tempo promoted by Louis Armstrong.

Idée fixe – A French expression for a transformable melody that recurs in every movement of a multi movement work.

Idiophone – A musical instrument that creates sound primarily using the vibration of the whole body, without the use of strings or membranes, such as the marimba, or a triangle.

Imitation – A polyphonic texture in which material is presented then echoed from voice to voice.

Impressionism – A modern French musical style based on blurred effects, beautiful tone colors, and fluid rhythms. It was promoted by Debussy around the turn of the 1900s.

Impromptu – Instrumental piece, usually for piano, that gives the impression of being improvised.

Improvisation – Instantaneous creation of music while it is being performed.

Inharmonic – A partial whose frequency is not an integer multiple of a given fundamental frequency.

Incidental music – A music genre performed during a theatrical play.

Instrumentation – The combination of instruments that a composition is written for.

Interlude – A short piece of music that serves to connect two major sections, usually mixing themes of both sections.

Intermezzo – A piece designed originally to be performed between the acts of a play or opera. Also, an interlude.

Interval – The measured distance between two musical pitches.

Inversion – A variation technique, in which the intervals of a melody are reversed. Also, a chord in which the bass note is not the root.

Jazz – A style of American modern popular music combining African and Western musical traits.

Jazz band – An instrumental ensemble comprised of woodwinds, that is, saxophones and clarinets, brasses, that is, trumpets and trombones, and rhythm section, that is piano or guitar, bass, and drum set.

Kettledrums – Similar to timpani.

Key – The central note, chord, or scale of a musical composition or movement. Also, a combination of a tonic and a mode.

Key signature – A series of sharps or flats written on a musical staff to indicate the key of a composition.

Keyboard instrument – Any instrument whose sound is initiated by pressing a series of keys with the fingers. The piano, the harpsichord, the organ, and the synthesizer are the most common types.

Koto – A Japanese plucked instrument with 13 strings and movable bridges.

Largo – A very slow, broad tempo.

Lead sheet – A piece of sheet music that contains a single staff for the melody, with the accompanying chords written above the staff.

Legato – A smooth, connected manner of performing a melody.

Leitmotif – A short musical signature tune associated with a person or concept in a Wagnerian Musikdrama.

Libretto – The sung or spoken text of an opera.

Lied – A German-texted art song, usually for one voice with piano accompaniment.

Loudness – A subjective experience that corresponds to sound intensity.

Lute – An ancient pear-shaped plucked instrument widely used in the Renaissance and Baroque eras.

Madrigal – A composition genre on a short secular poem, sung by a small group of unaccompanied singers, such as one on a part. The madrigal flourished in Italy from 1520 to 1610, and was adopted in England during the Elizabethan Age (c. 1600).

Major chord – A basic chord type, that includes the root, major third, and perfect fifth degrees of the major scale.

Major key – Music based on a major scale, which is traditionally considered happy sounding.

Major scale – A family of seven alphabetically ordered pitches within the distance of an octave, following an intervalic pattern matching the white keys from "C" to the one octave above "C" on a piano.

Marching band – A large ensemble of woodwinds, brass, and percussion used for entertainment at sporting events and parades, usually performing march-like music in a strong duple meter.

Mass – In music, a composition based on the five daily prayers of the Roman Catholic Mass Ordinary – Kyrie, Gloris, Credo, Sanctus, and Agnus Dei.

Mass Ordinary – The five daily prayers of the Catholic mass: *Kyrie, Gloris, Credo, Sanctus, Agnus Dei.*

Mass Proper – The approximately two dozen prayers of a mass that change each day to reflect the particular feast day of the liturgical calendar.

Marimba – A pitched percussion instrument comprised of wooden bars struck by mallets. It is a mellower version of the xylophone.

Mazurka – A type of Polish dance in triple meter, sometimes used by Chopin in his piano works.

Measure – A rhythmic grouping, set off in written music by a vertical barline. Also called a bar.

Mediant – The third degree of the major scale.

Medieval – A term used to describe things related to the Middle Ages (c.450–1450).

Melisma – A succession of many pitches sung while sustaining one syllable of text.

Melody – The musical element that deals with the horizontal presentation of pitch.

Membraphone – A musical instrument that relies on the vibration of a membrane, such as, a drum.

Metallophone – A musical instrument that consists of tuned metal bars which are struck to make sound. The glockenspiel and vibraphone are metallophones.

Meter – Beats organized into recurring and recognizable accent patterns (2/4, 3/4, 4/4, etc.).

Metronome – A mechanical, or electric, device that precisely measures tempo.

Mezzo – An Italian prefix that means medium.

Mezzo-forte – A medium loud dynamic marking (*F*).

Mezzo-piano – A medium quiet dynamic marking (*f*).

Mezzo-soprano – A dramatic woman's voice that combines the power of an alto with the primary high range of a soprano.

Microphonics – The interference caused by mechanical shock or vibration of elements in a system.

Microtone – A non-Western musical interval that is smaller than a Western half-step.

Middle Ages – An era dominated by Catholic sacred music (c.450–1450) which began as simple chant but grew in complexity in the 13th to 15th centuries by experiments in harmony and rhythm. The composers Pérotin and Machaut are representatives of that era.

MIDI – An acronym for musical instrument digital interface. It is an industrial standard protocol established in 1974, that allows digital synthesizers to communicate with computers.

Minimalism – A modern compositional approach promoted by Glass, Reich, and other composers, in which a short melodic, rhythmic, or harmonic idea is repeated and gradually transformed as the basis of an extended work.

Minor key – Music based on a minor scale, which is traditionally considered as sad sounding.

Minor scale – A family of seven alphabetically ordered pitches within the distance of an octave, following an intervalic pattern matching the white keys from "A" to an octave above "A" on a piano. It is identical do the *Aeolian* mode.

Minuet – An aristocratic dance in 3/4 meter.

Minuet and trio form – The traditional third-movement form of the Classic four-movement design, based on an aristocratic dance in 3/4 meter.

Mixer – A device that linearly combines input signals, in a desired proportion, to produce a resulting output signal.

Modal music – A type of composition based on one or mode modes.

Mode – A scale or key used in a musical composition. Major and minor are modes, as are ancient modal scales found in Western music before c.1680. Include the *Ionian*, *Dorian*, *Phrygian*, *Lydian*, *Mixolydian*, *Aeolian*, and *Locrian* modes.

Moderato – A moderate tempo.

Modern era – A musical era, from c.1890 to present, impacted by daring experimentation, advances in musical technology, and popular or non-Western influences. The main composers were Debussy, Schoenberg, Stravinsky, Copland, and Cage.

Modulation – The process of changing from one musical key to another.

Monophonic texture – A single-line texture with no harmony.

Motet – A polyphonic vocal piece set to a sacred Latin text that is not from the Roman Catholic mass.

Motive – Also called *motif*, is small musical fragment used to build a larger musical idea. It can be reworked in the course of a composition, as in the four-note motive in Beethoven's Symphony No. 5, in Do minor (C minor).

Mordent – An ornamental instruction to play a single note as a trill with an adjacent note.

Movement – A complete, independent division of a larger work.

Mp3 – A modern technology that allows digital sound to be compressed into files that are approximately eight times smaller than the original, with small loss of quality.

Musikdrama – A type of ultra-dramatic German operatic theater developed by Richard Wagner in the mid-to-late Romantic era.

Musique concrète – Music comprised of natural sounds that are recorded or manipulated electronically, or via magnetic tape. Also, a compositional approach promoted by Varése in the 1950s.

Mute – A device used to muffle the tone and volume of an instrument.

Natural scale – Diatonic scale.

Natural sign – A musical symbol that raises the pitch one half-step.

Nationalism – Musical styles that include folk songs, dances, legends, language, or other national imagery relating to a composer's native country.

Neo-Classicism – An early 20th century compositional style in which Classic forms and the esthetics of balance, clarity, and structural unity are combined with modern approaches to harmony, rhythm, and tone color.

New age – A style of popular music in the 1980s and 1990s, that rejected the hard-edged beat of rock music by focusing on nature sounds, sweet synthesized tone colors, acoustic instruments, and short hypnotically repetitive ideas.

Nocturne – A French for night piece, a type of character piece for solo piano that evokes the moods and images of night time.

Noise – An unwanted signal added to a received or measured one. It is a stochastic signal.

Non-metrical – Music without a regular beat or steady meter.

Non-Western music – Music from countries other than Europe and the Americas.

Notation – A system for writing music down so that critical aspects of its performance can be recreated accurately.

Note – A musical tone of a specific pitch. In music notation, a black or white oval-shaped symbol, with or without a stem or flag, that represents a specific rhythmic duration or pitch. It is called tone in American English.

Oboe – A nasal-sounding double-reed instrument that is the alto of the standard woodwind family.

Octave – A musical interval between two pitches in which the upper pitch vibrates twice as fast as the lower.

Opera – A large-scale, fully staged dramatic theatrical work involving solo singers, chorus, and orchestra.

Opera buffa – Comic Italian opera, usually in two acts.

Opera seria – Serious Italian opera, usually in three acts.

Oratorio – A large-scale sacred work for solo singers, chorus and orchestra that is not staged.

Orchestra – A large instrumental ensemble comprised of strings, woodwinds, brasses and percussion.

Orchestration – The technique of conceiving or arranging a composition for orchestra.

Ordinary – Short for Mass Ordinary.

Organ – A wind or keyboard instrument, usually with many sets of pipes controlled from two or more manuals, including a set of pedals played by the organist's feet. Also, a set of mechanical or electrical stops allow the player to open or close the flow of air to selected groups of pipes.

Organum – A type of early French medieval polyphony dating from (c.1000–1200), featuring a slow non metered chant in the lowest voice with one or more faster metrical voices sung above, in melismatic style. Also many notes sung on each syllable of text.

Oscillate – To move continually and regularly from one point to another.

Ostinato – A short rhythmic or melodic idea that is repeated exactly over and over throughout a musical section or work.

Overture – A one-movement orchestral introduction to an opera. Wagner, Bizet, and other composers, after 1850, use the term prelude instead to show dramatic unity between the overture and the theatrical drama that follows it.

Partials – Component frequencies, or individual sinusoids, that compose an instrumental tone.

Passion – Similar to the oratorio.

Pastorale – A piece written to imitate the music of shepherds. It is a tender melody usually written in moderate 6/8 or 12/8 time.

Pedal point – A note sustained below changing harmonies.

Pentatonic scale – A five-note scale, Also, a folk or non-Western scale having five different notes within the space of an octave.

Perfect cadence – A cadence that resolves from the dominant to the tonic.

Percussion instrument – An instrument on which sound is generated by striking its surface with an object.

Permutation – A type of melodic variation that completely rearranges the pitches of the original melody.

Phoneme – The shortest speech sound in language that distinguishes meaning.

Phon – A unit of loudness level for pure tones, equivalent to 1 dB at 1000 Hz (1 kHz). It indicates an individual's perception of loudness.

Phrase – A small musical unit, a sub-section of a melody, equivalent to a grammatical phrase in a sentence.

Pianissimo – A very quiet dynamic marking (*pp*).

Piano – A quiet dynamic marking (*p*).

Piano – A versatile modern keyboard instrument that makes sound via fingered keys that engage felt-tipped hammers that strike the strings. It was named pianoforte because it could be played softly or loudly.

Pianoforte – The original instrumental prototype of the piano, which appeared in the late Baroque or early Classic eras.

Pitch chroma – A term equivalent to pitch class, that refers to a note regardless of the octave in which it occurs.

Pitch – The relative highness or lowness of a musical sound, based on frequency of vibration. It is a psychological construct, related to the frequency of a tone and also to its position in the musical scale.

Pitch height – The psychological dimension of pitch, related to the sound intensity. Two tones may share the same pitch chroma, but differ in their perceived height.

Pizzicato – Refers to a type of violin playing in which a string is plucked by the fingers.

Plagal cadence – A chord progression in which the subdominant chord is followed by the tonic.

Polka – A lively Bohemian (Czech) dance, traditionally for the common classes.

Polonaise – A Polish nationalistic military dance used in some of Chopin's piano character pieces.

Polyphony – The mixing together of several simultaneous melodic lines. Also, a music with two or more sounds happening simultaneously.

Polyphonic texture – When two or more independent melodic lines are sounding at the same time.

Polyrhythm – When several independent rhythmic lines are sounding at the same time.

Polytonality – When music is played in two or more contrasting keys at the same time.

Postlude – A concluding section, usually at the end of a keyboard movement.

Precession time – The time required for a higher-frequency vibration to depart from and return into alignment with a lower-frequency vibration.

Prelude – A free-form introductory movement to a fugue or other more complex composition. Also a term used instead of overture, introduced by Wagner, Bizet and other later Romantic composers, to show dramatic unity between the introductory orchestral music and the theatrical drama that follows it.

Prepared piano – A modern technique, invented by John Cage, in which various natural objects (spoons, erasers, screws, etc.) are strategically inserted between the strings of a piano, in order to create unusual sounds.

Presto – A very fast tempo.

Program music – Instrumental music intended to tell a specific story, or set a specific mood or extra musical image.

Program symphony – A programmatic multi-movement work for orchestra.

Progression – A series of chords that functions similarly to a sentence or phrase in written language.

Propagation – Sound travel in a medium, usually subject to constraints but no impediment.

Proper – Short for Mass Proper.

Quadruple meter – A basic metrical pattern having four beats per measure.

Quarter note – A note of one beat's duration.

Quotation music – A composition extensively using quotations from earlier works, common since c.1960.

Raga – A melodic pattern used in the music of India.

Ragtime – A style of piano music developed around the turn of the 20th century, with a march-like tempo a syncopated right-hand melody, and a swing bass (oom-pah) left-hand accompaniment.

Range – The distance between the lowest and highest possible notes of an instrument or melody.

Rap – A style of popular music developed by Afro-Americans in the 1970s, in which the lyrics are spoken over rhythm tracks. Also called hip hop.

Rhapsody – A free fantasy, usually of heroic or national character, typically brilliant in style.

Recapitulation – The third aspect of Classic sonata form. In this section, both themes of the exposition are presented in the tonic key.

Resonance – The tendency of a system to vibrate at a certain frequency in response to energy induced at that frequency.

Resonant frequency – The frequency that is most effective to enable a vibrating system to return to its initial energy level by dissipation.

Restated – In the home key. The second theme gives up its opposing key center.

Recitative – A speech-like style of singing with a free rhythm over a sparse accompaniment.

Recorder – An ancient wooden flute.

Reed – A flexible strip of cane (or metal) that vibrates in the mouthpiece of a wind instrument.

Register – A specific coloristic portion of an instrumental or vocal range.

Renaissance – An era (c.1450–1600) that witnessed the rebirth of learning and exploration. This was reflected musically in a more personal style than seen in the Middle Ages. The important composers were Josquin des Prez, Palestrina, and Weelkes.

Requiem mass – A Roman Catholic mass for the dead.

Retrograde – A melody presented in backwards motion.

Retrograde inversion – A melody presented backwards and intervalically upside down.

Rhythm – The element of music as it unfolds in time. It refers to the duration of a series of notes, and to the way they group together into units.

Rhythm and blues – A style of Afro-American popular music that flourished in the 1940s and 1960s. It is a direct predecessor to rock and roll.

Riff – A short musical phrase, often repeated during the song. The repetitive motive of a song.

Ritardando – Gradually slowing down the tempo.

Ritornello form – A Baroque design that alternates big and small effects, as in *tutti versus solo*. Usually, the *tutti* section is a recurring melodic refrain.

Rock and roll – A style of popular music that emerged in the 1950s, as a result of the combination of Afro-American, Country-Western, and pop-music elements.

Romantic era – An era (c.1820–1890) of flamboyance, nationalism, the rise of superstar performers, and concerts, aimed at middle-class paying audiences. Orchestral, theatrical and soloistic music grew to spectacular heights of personal expression, and the main composers were Schubert, Berlioz, Chopin, Wagner, Brahms, and Tchaikovsky.

Rondo form – A Classic form in which a main melodic idea returns two or three times in alternation with other melodies (ABACA or ABACABA etc.).

Root note – The fundamental note in a chord. For instance, G (Sol) is the root note of the G major chord.

Rubato – A flexible approach to metered rhythm in which the tempo can be momentarily sped up or slowed down at will for greater personal expression.

Sackbut – An ancient brass instrument, ancestor to the trombone.

Saxophone – A family of woodwind instruments with a single reed and brass body. It is commonly used in jazz and marching band or concert band music.

Scale – A family of pitches arranged in an ascending or descending order. Also called *gamma*.

Scat singing – A style of improvised jazz singing sung on colorful nonsense syllables.

Scherzo – It is a short composition, or a movement from a larger work such as a symphony.

Scherzo and trio form – A musical movement based on a country dance in triple meter. It was replaced by the aristocratic minuet in the early 1800s as the usual third movement of the Classic four-movement design.

Score – Written notation that vertically aligns all instrumental/vocal parts used in a composition.

Sequence – The immediate repetition of a melodic passage on a higher or lower pitch level.

Set-theory – A method of harmonic analysis of clusters, for the atonal music that emerged from the set theory.

Serenade – A Classic instrumental chamber work similar to a small-scale symphony. It is usually performed for social entertainment of the upper classes.

Serialism – A method of modern composition in which the 12 chromatic pitches are put into a numerically-ordered series used to control various aspects of a work (melody, harmony, tone color, dynamics, instrumentation etc.).

Sforzando – A sudden stress on a note or chord (sfz).

Shakuhachi – A Japanese flute.

Shamisen – A banjo-like Japanese stringed instrument.

Sharp sign – A musical symbol that raises the pitch one half-step (#) or a semitone.

Shawm – An ancient double-reed woodwind instrument.

SI – International System of Units. The SI established the frequency of 440 Hz for the note La (A), in the equal temperament scale.

Signal – A visual, aural, or other form to convey information. Also, the information to be conveyed over, or a wave in a communication system.

Singspiel – A traditionally low-level type of comic light opera, featuring spoken German dialog interspersed with simple German songs.

Sitar – A long-necked stringed instrument of India.

Slash chord – Chord symbol indicating a specific bass note to play.

Snare drum – A non-pitched drum with two heads stretched over a metal shell. The lower head has metal wires strapped across it to produce a rattling sound.

Solfege – To sing the written notes, or melodies. It is also written as *solfeggio*.

Solo concerto – A three-movement work for a single soloist *versus* an orchestra.

Sonata – A Classic multi-movement work for a piano, or for one instrument with piano accompaniment.

Sonata form – The common first-movement form of Classic multi-movement instrumental works. Essentially a musical debate between two opposing key centers characterized by three dramatic structural divisions within a single movement: exposition (two opposing keys are presented), development (harmonically restless), and recapitulation (all material is presented in the home key). Also called sonata-allegro form.

Sonata-rondo form – A formal design that combines aspects of sonata form and rondo form: an ABACABA design in which the opening ABA is the exposition (two opposing keys presented in "A" *versus* "BA"); C is the development (harmonically restless); and the last ABA is the recapitulation (all material is presented in the home key).

Sone – A unit of loudness, that indicates how loud a sound is perceived by the human ear.

Song – A small-scale musical work that is sung. A German song is a called *Lied*, a French song is a *chanson*, and an Italian song is a *canzona*.

Song cycle – A set of poetically unified songs (for one singer accompanied by either piano or orchestra).

Soprano – The highest ranged woman's voice, or a high pre-pubescent boy's voice. Also the highest-sounding instrument of an instrumental family.

Sousaphone – An ultra-bass brass instrument designed for use in marching bands.

Sprechstimme – A half-spoken, half-sung style of singing on approximate pitches, developed by Schoenberg in the early 1900s.

Staccato – Short, detached notes.

Staff – A set of five horizontal lines used in musical notation to indicate note names.

String instrument – An instrument that is played by placing one's hands directly on the strings, such as violin, viola, cello, double bass, harp, guitar, dulcimer, psaltery, and the ancient viols.

String quartet – A chamber ensemble of two violins, viola and cello, devised in the early Classic era. Also a multi-movement work (genre) for two violins, viola and cello.

Strophic form – A song form featuring several successive verses of text sung to the same music.

Subdominant – The fourth chord or tone of a scale (IV).

Subject – The main melodic idea of a fugue. A *motif*, phrase, or melody that is the basic element in a musical composition.

Submediant – The sixth chord or tone of a scale, also called superdominant.

Suite – A collection of dance movements.

Suspension – A dissonant note used within a chord to create tension. The suspended note is usually the fourth of the chord, which then resolves down to the third.

Swing – A term to describe Big Band jazz music of the 1930s and 1950s.

Symphonic poem – A single-movement programmatic work for orchestra.

Symphony – A multi-movement work for orchestra.

Syncopation – An off-the-beat accent. A temporary displacement of the regular metrical accent in music caused typically by stressing the weak beat.

Synthesizer – A modern electronic keyboard instrument capable of generating a multitude of sounds.

Tabla – A pair of drums used to accompany the music of India.

Tablature – A simplified system of notation used for the guitar, that dates back to the Renaissance period. Also, TAB is a graphical system that guitar players use for reading numbers instead of notes.

Tala – A rhythmic pattern used in the music of India.

Tempo – The speed or pace of the musical beat. Tempo is measured in beats per minute (BPM).

Tenor – A high-ranged male voice.

Ternary form – The usual ABA design, that indicates statement, contrast, and restatement.

Tessitura – The range of pitches used in a composition, in an instrument, or in a voice.

Texture – The element focusing on the number of simultaneous musical lines being sounded.

Theme – The main self-contained melody of a musical composition.

Theme and variations form – A theme is stated then undergoes a series of sectional alterations.

Through-composed form – A song form with no large-scale musical repetition.

Tie – A curved line connecting two notes, an indication to hold the tone for the combined rhythmic value of both notes.

Timbre – Another term for tone color.

Time signature – Also known as meter signature or measure signature, is a notational convention used to specify how many beats, or pulses, are contained in each measure, or bar, and which note value is equivalent to a beat.

Timpani – Various-sized kettle-shaped pitched drums. Also, a tenor instrument of the percussion family.

Tone – A sound played or sung at a specific pitch. The tone also involves a duration, a certain intensity, and the instrument or voice timbre.

Tone color – The unique, characteristic sound of a musical instrument or voice.

Tone cluster – A modern technique of extreme harmonic dissonance created by a large block of pitches sounding simultaneously.

Tonality – Music centered around a home key, based on a major or minor scale.

Tone row – An ordered series of 12 chromatic pitches used in serialism.

Tonic – The first note or chord, of a scale or key built on that degree (I).

Tonic key – The home key of a tonal composition.

Transition – A bridge section between two musical ideas.

Transposing instrument – An instrument that is not notated at its sounding pitch.

Transposition – Shifting a piece to a different pitch level, also changing the key of a melody while keeping the same intervalic relationship.

Treble clef – The G clef, also a symbol that indicates it.

Treble boost – An accentuation of higher audio frequencies in the response of an equipment.

Tremolo – Rapid repetition of a pitch, that is, bowing a string rapidly while maintaining a constant pitch.

Triad – A three-note chord built on alternating scales steps.

Trill – A rapid alternation of two close pitches to create a shaking ornament on a melodic note.

Trio sonata – A Baroque multi-movement chamber work for four performers, involving two violins and basso continuo.

Triple meter – A common meter with three beats per measure.

Triplet – A rhythmic grouping of three equal-valued notes played in the space of two, that is indicated in written music by the symbol "3" above the grouping.

Trombone – A family of brass instruments that change pitch via a movable slide (alto, tenor, and bass versions are common).

Trumpet – A valved instrument that is the soprano of the modern brass family.

Tuba – A large valved brass instrument. Also, the bass of the modern brass family.

Tubular bells – Musical instruments in the percussion family. Their sound resembles that of church bells, carillon, or a bell tower. Also called chimes.

Tutti – Italian word for all or everyone. It is an indication for all performers to play together.

Ud – A lute-like, pear-shaped, fretless stringed instrument commonly used in music from the Middle East.

Unison – The rendering of a single melodic line by several performers simultaneously.

Upbeat – The weak beat that comes before the strong downbeat of a musical measure.

Variation – The compositional process of changing some aspects of a musical work, while retaining others.

Verismo – A style of true-to-life Italian opera that flourished at the turn of the 20th century.

Verse – In popular music, the first section of a song, preceding the chorus.

Vibrato – Small fluctuations in pitch used to make a sound more expressive.

Viol – An ancient string instrument, that is an ancestor to the modern violin.

Viol' da gamba – A Renaissance bowed string instrument held between the legs like a modern cello.

Viola – The alto instrument of the modern string family.

Violin – The soprano instrument of the modern string family.

Violoncello – The full name of the cello. It is the tenor instrument of the modern string family.

Virtuoso – A performer of extraordinary ability.

Vivace – A lively tempo.

Volume – The relative quietness or loudness of an electrical impulse.

Waltz – An aristocratic ballroom dance in triple meter that flourished in the Romantic period.

Whole step – An interval twice as large as a half-step. Example: the distance between C and D on a piano.

Wolf note – A note, on any type of acoustic instrument, that is markedly different in tone or quality to the others. Also, a note that sounds weak or irregular because of the properties of acoustic resonance.

Whole-tone scale – A scale made of six whole steps that avoids any sense of tonality. Example: C D E F# G# A#.

Woodwind instrument – An instrument that produces its sound from a column of air vibrating within a multi-holed tube.

Word painting – In vocal music, musical gestures that reflect the specific meaning of words. It is a common aspect of the Renaissance madrigal.

World beat – The collective term for contemporary popular third-world musical styles. Also called ethno pop.

Xylophone – A pitched percussion instrument consisting of flat wooden bars on a metal frame that are struck by hard mallets.

Yoruba music – A musical genre of the Yoruba people of Nigeria, Benin, and Togo.

Zabumba – A zabumba is a type of bass drum used in Brazilian music.

Bibliography

Abad, F., *Do Re Qué? – Guia Prática de Iniciación al Lenguage Musical.* Berenice Manuales, Madrid, España (2006).

Abromont, C. and de Montalem bert, E., *Teoría de la Música – Una Guia.* Fondo de Cultura Económica, México D. F., México (2010).

Alencar, M. S., *Principles of Communications (Portuguese).* University Publishers, UFPB, João Pessoa, Brazil (1999).

Alencar, M. S., **A Análise de Fourier sobre o Aquecimento Global I.** Artigo para jornal eletrônico na Internet, Jornal do Commercio *On Line*, Recife, Brasil (2007a).

Alencar, M. S., **A Análise de Fourier sobre o Aquecimento Global II.** Artigo para jornal eletrônico na Internet, Jornal do Commercio *On Line*, Recife, Brasil (2007b).

Alencar, M. S., **A Análise de Fourier sobre o Aquecimento Global III.** Artigo para jornal eletrônico na Internet, Jornal do Commercio *On Line*, Recife, Brasil (2007c).

Alencar, M. S., **A Análise de Fourier sobre o Aquecimento Global IV.** Artigo para jornal eletrônico na Internet, Jornal do Commercio *On Line*, Recife, Brasil (2007d).

Alencar, M. S., **A Natureza Quântica da Música.** Artigo para jornal eletrônico na Internet, NE10 – Sistema Jornal do Commercio de Comunicação, Recife, Brasil (2018a).

Alencar, M. S., **A Percepção do Som.** Artigo para jornal eletrônico na Internet, NE10 – Sistema Jornal do Commercio de Comunicação, Recife, Brasil (2018b).

Alencar, M. S., **Feitos de Silêncio e Som.** Artigo para jornal eletrônico na Internet, NE10 – Sistema Jornal do Commercio de Comunicação, Recife, Brasil (2018c).

Alencar, M. S., **O Material e o Virtual.** Artigo para jornal eletrônico na Internet, NE10 – Sistema Jornal do Commercio de Comunicação, Recife, Brasil (2018d).

Alencar, M. S., **O Som Organizado.** Artigo para jornal eletrônico na Internet, NE10 – Sistema Jornal do Commercio de Comunicação, Recife, Brasil (2018e).

Alencar, M. S., **O Áudio e o Vídeo.** Artigo para jornal eletrônico na Internet, NE10 – Sistema Jornal do Commercio de Comunicação, Recife, Brasil (2018f).

Alencar, M. S. and da Rocha Jr., V. C., *Communication Systems.* Springer, ISBN 0-387-25481-1, Boston, USA (2005).

Alipour, F., Berry, D. A. and Titze, I. R., **A Finite-Element Model of Vocal-Fold Vibration.** *The Journal of the Acoustical Society of America* (2000).

Almada, C., *Harmonia Funcional.* Editora da Unicamp, Campinas, Brazil (2009).

Anumanchipalli, G. K., Cheng, Y. C., Fernandez, J., Huang, X., Mao, Q. and Black, A. W., **KlaTTStat: Knowledge-based Statistical Parametric Speech Synthesis.** *7th ISCA Workshop on Speech Synthesis* (2010).

Araújo, F. P. O., **Imitação da Voz Humana através do Processo de Análise-por-Síntese utilizando Algoritmo Genético e Sintetizador de Voz por Formantes.** Tese de doutorado, Universidade Federal de Santa Catarina (2015).

Arbonés, J. and Milrud, P., *Música y Matemáticas.* National Geografic – RBA Libros, Barcelona, España (2012).

Aronson, A. E. and Bless, D. M., **Clinical Voice Disorders.** *Thieme Medical Publishers* (2009).

Ashton, A., *Harmograph: A Visual Guide to the Mathematics of Music.* Bloomsbury Publishing Plc, New York, USA (2001).

Ball, P., *L'Instinto Musicale – Como e Perché Abbiamo la Musica Dentro.* Edizione Dedalo, Viale Luigi Javobini, Bari, Italia (2010).

Baskakov, S. I., *Signals and Circuits.* Mir Publishers, Moscow, USSR (1986).

Behlau, M. and Rehder, M. I., **Higiene Vocal para o Canto Coral.** *Editora Revinter* (1997).

Benade, A. H., *Fundamentals of Musical Acoustics.* Dover Publications, Inc., New York, USA (1990).

Bennet, R., *Uma Breve História da Música.* Editora Zahar, Rio de Janeiro, Brazil (1986).

Bennett, R., *Elementos Básicos da Música.* Jorge Zahar Editor Ltda., Rio de Janeiro, Brazil (1998).

Berry, D. A., Herzel, H., Titze, I. R. and Krischer, K., **Interpretation of Biomechanical Simulations of Normal and Chaotic Vocal Fold Oscillations with Empirical Eigenfunctions.** *Journal of the Acoustical Society of America*, 95(6):3595–3604 (1994).

Berry, D. A. and Titze, I. R., **Normal Modes in a Continuum Model of Vocal Fold Tissues.** *Journal of the Acoustical Society of America*, 100(5):3345–3354 (1996).

Bessette, B., Salami, R., Lefebvre, R., Jelínek, M., Rotola-Pukkila, J., Vainio, J., Mikkola, H. and Järvinen, K., **The Adaptive Multirate Wideband Speech Codec (AMR-WB).** *IEEE Transactions on Speech and Audio Processing*, 10(8):620 – 636 (2002).

Brandão, A. S., **Modelagem Acústica da Produção da Voz Utilizando Técnicas de Visualização de Imagens Médicas Associadas a Métodos Numéricos.** Tese de doutorado, Universidade Federal Fluminense, Niterói (2011).

Brito, J., **Genetic Learning of Vocal Tract Area Functions for Articulatory Synthesis of Spanish Vowels.** *Applied Soft Computing*, pp. 1035–1043 (2007).

Buck, J., *Play Guitar in 10 Easy Lessons.* Hachete UK Company, London, Great Britain (2014).

Burrows, T., *Total Piano Tutor – The Ultimate Guide to Learnig and Mastering the Piano.* Sevenoaks, Carlton Books Limited, London, United Kingdom (2002).

Burrows, T., *How to Read Music – Reading Music Made Simple.* St. Martin's Press, New York, USA (2004).

Capone, P., *Learn to Play Guitar: A Beginner's Guide to Playing Acoustic and Electric Guitar.* Chartwell Bookd, Inc., New York, USA (2007).

Capone, P., *Learn to Play Bass Guitar: A Beginner's Guide to Bass Guitar.* Chartwell Bookd, Inc., New York, USA (2009).

Carlson, B. A., *Communication Systems.* McGraw-Hill, Tokyo, Japan (1975).

Castro, R., *Bossa Nova: The Story of the Brazilian Music That Seduced the World.* A Capella Books, Chicago Review Press, Inc., Chicago, USA (2000).

Cebolo, E. A., *Teoria Mágica Musical.* Musicarte, Porto, Portugal (2015).

Chapman, R., *The New Complete Guitarist – The All-visual Approach to Mastering the Guitar.* Dorling Kindersley Ltd., London, United Kingdom (2003).

Chediak, A., *Dicionário de Acordes Cifrados.* Irmãos Vitale S/A Ind. Com., Rio de Janeiro, Brazil (1984).

Chediak, A., *Harmonia & Improvisação.* Lumiar Editora, Rio de Janeiro, Brazil (1986).

Cook, P. R., *Real Sound Synthesis for Interactive Applications.* A. K. Peters, CRC Press, Natick, USA (2003).

Costa, W. C. A., **Reconhecimento de Fala Utilizando Modelos de Markov Escondidos (HMM's) de Densidades Contínuas.** Dissertação de mestrado, Universidade Federal da Paraíba (1994).

Costa-Neto, M. L., **Um Modelo para Geração de Prosódia de Palavras em Conversores Texto-Fala para a Língua Portuguesa Falada no Brasil.** Tese de doutorado, Universidade Federal de Campina Grande (2004).

Crovato, C. D. P., **Classificação de Sinais de Voz Utilizando a Transformada Wavelet Packet e Redes Neurais Artificiais.** Dissertação de mestrado, Faculdade de Engenharia da Universidade do Porto (2004).

Curtu, I., Stanciu, M. D., Cretu, N. C. and Rosca, I. C., **"Modal Analysis of Different Types of Classical Guitar Bodies".** In *Proceedings of the 10th WSEAS International Conference on Acoustics & Music: Theory and Applications*, pp. 30–35, Prague, Czech Republic. World Scientific and Engineering Academy and Society (WSEAS) (2009).

da Cunha, N. P., *Matemática & Música: Diálogo Interdisciplinar, 2nd edition.* Editora Universitária da UFPE, Recife, Brazil (2008).

da Silva, K. W. A. X., **Sistema de Conversão Texto-Fala com Busca Otimizada de Unidades Acústicas em Banco de Voz.** Dissertação de mestrado, Universidade Federal do Rio de Janeiro (2011).

Davenport, W. B. and Root, W. L., *An Introduction to the Theory of Random Signals and Noise.* Wiley-IEEE Press, New York, USA (1987).

Davis, D. and Davis, C., *Sound system Engineering.* Howard W. Sams & Co., Carmel, USA (1987).

de la Fuente, J. M. M., *Las Vibraciones de la Música.* Editorial Club Universitario, San Vicente, España (2014).

de Lacerda, O. C., *Compêndio de Teoria Elementar da Música, 15th edition.* Ricordi Brasileira S/A, São Paulo, Brazil (1966).

de Mattos Priolli, M. L., *Princípios Básicos de Música para a Juventude, 33rd edition*, volume 2. Casa Oliveira de Músicas Ltda., Rio de Janeiro, Brazil (2013a).

de Mattos Priolli, M. L., *Princípios Básicos de Música para a Juventude, 54th edition*, volume 1. Casa Oliveira de Músicas Ltda., Rio de Janeiro, Brazil (2013b).

de Paula, C. A., **A música no ensino médio da escola pública do município de curitiba: Aproximações e proposições conceituais à realidade concreta.** Dissertação de mestrado, Universidade Federal do Paraná, Curitiba, Brazil (2007).

Degottex, G., **Glottal Source and Vocal-Tract Separation. Estimation of Glottal Parameters, Voice Transformation and Synthesis using a Glottal Model.** Tese de doutorado, Université Paris (2010).

Deller, J. R., Proakis, J. G. and Hansen, J. H. L., *Discrete-Time Processing of Speech Signals.* Prentice-Hall (1993).

Dias, S. O., **Estimation of the Glottal Pulse from Speech or Singing Voice.** Master's thesis, School of Engineering of the University of Porto (2012).

do N. C. Costa, S. L., **Análise Acústica, Baseada no Modelo Linear de Produção da Fala, para Discriminação de Vozes Patológicas.** Tese de doutorado, Universidade Federal de Campina Grande (2008).

Doering, E., *Musical Signal Processing with LabVIEW – Subtractive Synthesis.* Connexions, Rice University, Houston, USA (2012).

dos Santos, R. J. F., **Avaliação de Pacientes com Paralisia Unilateral das Pregas Vocais.** Dissertação de mestrado, Universidade de Aveiro (2009).

du Sautoy, M., *A Música dos Números Primos: a História de um Problema não Resolvido na Matemática.* Jorge Zahar Editor Ltda., Rio de Janeiro, Brasil (2007).

Dutoit, T., *A Introduction to Text-to-Speech Synthesis.* Academic Publishers (2011).

Cataldo, E., Sampaio, R. and Nicolato, L., **Uma Discussão sobre Modelos Mecânicos de Laringe para Síntese de Vogais.** *Engevista*, 6(1):47–57 (2004).

Eendebak, B., **Design of a Classical Guitar.** [Online; accessed October 23, 2019] (2019).

Einstein, A., **Näherungsweise Integration der Feldgleichungen der Gravitation.** *Sitzungsberichte der Königlich Preussischen Akademie der Wissenschaften Berlin*, XLVII(8):688–696 (1916).

Elejabarrieta, M. J., Ezcurra, A. and Santamaria, C. M., **"Coupled Modes of the Resonance Box of the Guitar".** *The Journal of the Acoustical Society of America*, 113(5):2283–2292 (2002).

Encina, M., Yuz, J., Zañartu, M. and Galindo, G., **Voice fold modeling through the port-hamiltonian systems approach**. *IEEE Conference on Control Applications*, pp. 1558–1563 (2015).

Eves, H., *Introdução à História da Matemática.* Editora da Unicamp, Campinas, Brasil (2011).

Fant, G., **Vocal-Source Analysis – A Progress Report.** *TL-QPSR*, pp. 31–53 (1979).

Fant, G., Liljencrants, J. and guaq Lin, Q., **A Four Parameter Model of Glottal Flow.** *TL-QPSR*, pp. 1–13 (1985).

Fechine, J. M., **Verificação de locutor utilizando modelos de markov escondidos (HMMs) de densidades discretas**. *Dissertação de Mestrado, Universidade Federal da Paraíba, Departamento de Engenharia Elétrica* (1994).

Fernandes, P. A. F. P., **Modelo do Sistema de Produção de Voz Aplicável à Detecção de Anomalias nas Cordas Vocais.** Dissertação de mestrado, Faculdade de Engenharia da Universidade do Porto (2004).

Flanagan, J. and Landgraf, L., **Self-Oscillating Source for Vocal-Tract Synthesizers.** *IEEE Transactions on Audio and Eletroacoustics*, 16(1):57–64 (1968).

Franceschina, J., *Music Theory Through Musical Theatre.* Oxford University Press, New York, USA (2015).

Friedland, E., *Bass Method: Complete Edition.* Hal Leonard Corporation, Milwaukee, USA (2004).

Furui, S., *Digital Speech Processing, Synthesis, and Recognition.* Tokai University Press (1985).

G.729, I.-T. R., **General Aspects of Digital Transmission Systems Terminal Equipments – Coding of Speech at 8 kbits/s Using Conjugate-Structure Algebraic-Code-Excited Linear-Prediction (CS-ACELP)** (1996).

Gagliardi, R. M., *Introduction to Communications Engineering.* Wiley, New York (1988).

Gobl, C., **The Voice Source in Speech Communication.** Doctoral thesis, School of Engineering of the University of Porto (2003).

Gradshteyn, I. S. and Ryzhik, I. M., *Table of Integrals, Series, and Products.* Academic Press, Inc., San Diego, California (1990).

Greiss, R., Rocha, J. and Matida, E., **Modal Analysis of a Parameterized Model of Pathological Vocal Fold Vibration.** *IEEE EMBS International Student Conference*, pp. 1–4 (2016).

Grout, D., *A History of Western Music (3rd edition).* W. W. Norton & Company, New York, USA (1980).

Hammer, J., *Absolute Beginners Keyboard – The Complete Picture Guide to Playing Keyboard.* MidPoint Press, London, United Kingdom (1999).

Hauer, J. M., **Schriften, Manifeste, Dokumente.** Dvd-rom, ISBN 978-3-85151-076-8, Lafite, Wien (2007).

Haykin, S., *Communication Systems.* Wiley Eastern Limited, New Delhi, India (1987).

Haykin, S., *Digital Communications.* John Wiley and Sons, New York (1988).

Henrique, L. L., *Acústica Musical, 5th edition.* Fundação Calouste Gulbenkian, Lisboa, Portugal (2014).

Henry, S., *Play Piano – A First Book for Beginners of All Ages.* Bookmark Limited, The Brown Reference Group plc, London, United Kingdom (2003).

Hodge, D., *Playing Bass Guitar.* Alpha Books, New York, USA (2006).

Howard, I. S. and Huckvale, M. A., **Training a Vocal Tract Synthesizer to Imitate Speech Using Distal Learning.** *Proceedings of InterSpeech* (2005).

Howard, J., *Apredendo a Compor.* Jorge Zahar Editor Ltda., Rio de janeiro, Brazil (1991).

Hsu, H. P., *Fourier Analysis (Portuguese).* Livros Técnicos e Científicos Publishers Ltd., Rio de Janeiro, Brasil (1973).

Hüttner, B., Döllinger, M., Luegmair, G., Eysholdt, U., Ziethe, A. and Gurlek, E., **Parameter Optimization for a Time-Dependent Multi-Mass Model for the Pharyngo-Esophageal Segment.** *Proceedings of Models and Analysis of Vocal Emissions for Biomedical Applications* (2011).

Ishizaka, K. and Flanagan, J., **Synthesis of Voiced Sounds from Two-Mass Model of the Vocal Cords.** *Bell System Technical Journal*, 51:1233–1268 (1972).

Johnson, T., *Self-Similar Melodies.* Editions 75, Paris, France (1996).

Johnston, I., *Measured Tones.* IOP Publishing Ltd., Bristol, United Kingdom (1989).

Joseph E. Hawkins, **Human Ear.** [Online; accessed May 29, 2019] (2019).

Juanilla, J. M., *Armonía Moderna y Otras Herramientas Compositivas.* Musikastist 31 y ZAKstudio, Madrid, Spain (2014).

Kennedy, M., Kennedy, J. B. and (Editor), T. R.-J., *The Oxford Dictionary of Music.* Oxford University Press, Inc., Oxford, United Kingdom (2013).

Kent, R. D. and Read, C., **The Acoustics Analysis of Speech.** *Singular Publishing Group Inc.* (1992).

Klabbers, E. A. M., **Segmental and Prosodic Improvements to Speech Generation.** Tese de doutorado, Technische Universiteit Eindhoven, Netherlands (2000).

Klatt, D., **Software for a Cascade/Parallel Formant Synthesizer.** *Journal of the Acoustical Society of America* (1980).

Klatt, D. and Klatt, L., **Analysis, Synthesis, and Perception of Voice Quality Variations Among Female and Male Speakers.** *Journal of the Acoustical Society of America* (1990a).

Klatt, D. and Klatt, L., **Analysis, Synthesis, and Perception of Voice Quality Variations among Female and Male Talkers.** *Jurnal of the Acoustical Society of America*, pp. 820–857 (1990b).

Knopp, K., *Theory and Application of Infinite Series.* Dover Publications, Inc., New York (1990).

Károlyi, O., *Introdução à Música.* Martins Fontes, São Paulo, Brasil (2002).

Lathi, B. P., *Modern Digital and Analog Communication Systems.* Holt, Rinehart and Winston, Inc., Philadelphia, USA (1989).

Latsch, V. L., **Construção de um Banco de Unidades para Síntese da Fala por Concatenação no Domínio Temporal.** Dissertação de mestrado, Universidade Federal do Rio de Janeiro (2005).

Levitin, D. J., *This is Your Brain in Music: The Science of a Human Obsession.* Plume, Penguin Group, New York, USA (2006).

Liénard, J.-S., *Les Processus de la Communication Parlée.* Masson, Paris, France (1977).

Loy, D. G., *Musimathics: The Mathematical Foundations of Music*, volume 2. The MIT Press, Cambridge, USA (2011a).

Loy, D. G., *Musimathics: The Mathematical Foundations of Music*, volume 1. The MIT Press, Cambridge, USA (2011b).

Lucero, J. C., **Oscillation Hysteresis in a Two-Mass Model of the Vocal Folds.** *Journal of Sound and Vibration*, 282(3–5):1247–1254 (2005).

Lyra, C., **World Music Instrument: The Berimbau.** Internet site: https://centerforworldmusic.org/2015/06/world-music-instruments-the-berimbau/, Center for World Music, San Diego, USA (2019).

Dajer, M. E., **Padrões Visuais de Sinais de Voz através de Técnica de Análise de Não-Linear.** Dissertação de mestrado, Escola de Engenharia de São Carlos (2006).

Maia, R. S., **Codificação CELP e Análise Espectral da Voz.** Dissertação de mestrado, Universidade Federal do Rio de Janeiro (2000).

Maor, E., *La Música e los Números – De Pitágoras a Schoenberg.* Turner Publicaciones S. L., Madrid, España (2018).

Margolis, E. and Eldar, Y. C., **"Nonuniform Sampling of Periodic Bandlimited Signals".** *IEEE Transactions on Signal Processing*, 56(7):2728–2745 (2008).

Marinus, J. V. M. L., Araújo, J. M. F. R., Gomes, H. M. and Costa, S. C., **On the Use of Cepstral Coefficients, Multilayer Perceptron Networks and Gaussian Mixture Models for Vocal Fold Edema diagnosis.** *Biosignals and Biorobotics Conference.*, pp. 1–6 (2013).

Martineau, J., *The Elements of Music: Melody, Rhythm, & Harmony.* Wooden Books Ltd, Glastonbury, UK (2008).

Mascarenhas, M., *Método Rápido para Tocar Teclado, Primeiro Volume.* Irmãos Vitale S/A Ind. e Com., Rio de Janeiro, Brasil (1991).

Med, B., *Solfejo, Terceira Edição.* MusiMed Edições Musicais, Importação e Exportação, Brasília, Brazil (1996a).

Med, B., *Teoria da Música, 4th edition.* MusiMed Edições Musicais, Importação e Exportação, Brasília, Brazil (1996b).

Meireles, A., *Análise Acústica da Fala.* Editora Cortez (2015).

Mendes, M. F., *Arranjando Frevo de Rua.* Cepe Editora, Recife, Brazil (2017).

Mendes, M. F., *Arranjando Frevo-Canção.* Cepe Editora, Recife, Brazil (2019).

Mendez, A., Garcia, B., Vicente, J., Ruiz, I. and Sanchez, K., **Objective Model of Vocal Folds, Based on Glottal Closure, Opening Angles and Morphologic Criteria.** *Signal Processing and Its Applications.*, pp. 1–4 (2007).

Miller, M., *The Complete Idiot's Guide to Music Composition.* Alpha, Penguin Random Hause LLC, New York, USA (2005).

Moon, E., *La Guitarre – C'est Pas Sorcier.* Hachette Livre, Départment Marabout, Rua Jean Bleuzen, Vanves, France (2015).

Morais, E. S., **Algoritmo OPWI e LDM-GA para Sistemas de Conversão Texto-Fala de Alta Qualidade Empregando a Tecnologia SCAUS.** Tese de doutorado, Unicamp (2006).

Moulton, R., **Arranging Music for A Cappella.** Internet site, WorldPress.com, moultano.wordpress.com/2006/04/10/arranging-music-for-a-cappella-3kbzhsxyg4467-2/ (2019).

Neely, B., *Piano for Dummies.* Wiley Publishing, Inc., New York, USA (1988).

Newman, J. R., *The World of Mathematics*, volume 4. Dover Publications, Inc., New York, USA (2000a).

Newman, J. R., *The World of Mathematics*, volume 1. Dover Publications, Inc., New York, USA (2000b).

Nikolaidis, R., **A generative model of tonal tension and its application in dynamic realtime sonification**. Master of science in music technology in the school of department of music, Georgia Institute of Technology, Atlanta, USA (2011).

Oberhettinger, F., *Tables of Fourier Transforms and Fourier Transforms of Distributions.* Springer-Verlag, Berlin (1990).

Pacheco, F. S., **Técnicas de Processamento de Sinais para Alteração de Parâmetros Prosódicos Aplicadas a um Sistema de Conversão Texto-Fala para a Língua Portuguesa Falada no Brasil.** Dissertação de mestrado, Universidade Federal de Santa Catarina (2001).

Palmer, W. A., Manus, M. and Lethco, A. V., *Scale, Chords, Arpeggios & Cadences.* Alfred Publishing Co. Inc., USA (1994).

Papoulis, A., *Signal Analysis.* McGraw-Hill, Tokyo (1983).

Paranaguá, E. D. S., **Segmentação Automática do Sinal de Voz Para Sistemas de Conversão Texto-Fala.** Tese de doutorado, Universidade Federal do Rio de Janeiro (2012).

Patel, A. D., *Music, Language, and the Brain.* Oxford University Press, New York, USA (2010).

Patel, T. B. and Patil, H. A., **Novel Approach for Estimating Length of the Vocal Folds using Fujisaki Model.** *Chinese Spoken Language Processing.*, pp. 308–312 (2014).

Pen, R., *Introduction to Music.* McGraw-Hill, New York, USA (1992).

Pesic, P., **Euler's Musical Mathematics.** *Years Ago*, 35(2):35–39 (2013).

Pfeiffer, P., *Bass Guitar for Dummies.* Wiley Publishing, Inc., Indianopolis, USA (2010).

Philippsen, A., Reinhart, F. R. and Wrede, B., **Learning how to Speak: Imitation-based Refinement of Syllable Production in an Articulatory-Acoustic Model.** *IEEE Int. Conf. on Development and Learning and on Epigenetic Robotics (ICDL)*, pp. 187–192 (2014).

Pilhofer, M. and Day, H., *Teoria Musical para Leigos.* Alta Books, Rio de Janeiro, Brasil (2013).

Powell, J., *How Music Works: The Science and Psycology of Beautiful Sounds, from Beethoven to the Beatles and Beyond.* Little, Brown and Company, New York, USA (2010).

Rabiner, L. R. and Juang, B., *Fundamentals on Speech Recognition.* Prentice Hall (1996).

Rabiner, L. R. and Schafer, R. W., *Digital Processing of Speech Signals.* Prentice Hall, New Jersey (1978).

Rocha, R. B., *Modelo de Produção de Voz Baseado na Biofísica da Fonação.* PhD thesis, Coordenação de Pós-Graduação em Engenharia Elétrica, Universidade Federal de Campina Grande, Campina Grande, Brasil (2017).

Rodriguez, J. M. L., *Breve História de la Música.* Ediciones Nowtilus, S. L., Madrid, España (2017).

Roederer, J. G., *Introdução à Física e Psicofísica da Música.* Editora da Universidade de são Paulo, São Pauo, Brazil (1998).

Rosa, M. O., **Laringe Digital.** Tese de doutorado, Universidade de São Paulo (2002).

Sacks, O., *Musicophilia: Tales of Music and the Brain.* Random House, Inc., New York, USA (2008).

Salami, R., Lefebvre, R., Lakaniemi, A., Kontola, K., Bruhn, S. and Taleb, A., **Extended AMR-WB for High-Quality Audio on Mobile Devices.** *IEEE Communications Magazine*, pp. 90 – 97 (2006).

Sallet, P., *Musijeunes – Classe de Sixième.* Connaissance de la Musique – S.E.D.P. Editeur, Roma, Italia (1974).

Sallet, P., *Musijeunes.* Connaissance de la Musique – S.E.D.P. Editeur, Roma, Italia (1976).

Schonbrun, M., *The Everything Essential Music Theory Book.* Adms Media, F+W Media, Inc., Avonm USA (2014).

Schwartz, M. and Shaw, L., *Signal Processing: Discrete Spectral Analysis, Detection, and Estimation.* McGraw-Hill, Tokyo (1975).

Selmini, A. M., **Sistema Baseado em Regras para o Refinamento da Segmentação Automática de Fala.** Tese de doutorado, Universidade Estadual de Campinas, Campinas, Brasil (2008).

Shepheard, G., *Get Started in Playing Piano.* Hodder & Stoughton, London, Great Britain (2014).

Silva, E. L. F., **Estimativas de Comportamento Vocálico de Locutores e um Novo Sistema de Separação Silábica.** Dissertação de mestrado, Universidade Federal de Pernambuco (2012).

Silva, L. M., **Contribuições para a Melhoria da Codificação CELP a Baixas Taxas de Bits.** Tese de doutorado, Pontifícea Universidade Católica do Rio de Janeiro (1996).

Simões, F. O., **Implementação de um Sistema de Conversão Texto-Fala para o Português do Brasil.** Tese de mestrado, Unicamp (1999).

Sándor, J., **Euler and Music. A Forgotten Arithmetic Function by Euler.** *Octogon Mathematical Magazine*, 17(1):265–271 (2009).

Sotero, R. F. B., **Novas Abordagens para Codificação de Voz e Reconhecimento Automático de Locutor Projetadas Via Mascaramento Pleno em Frequência por Oitava.** Dissertação de mestrado, Universidade Federal de Pernambuco (2009).

Spiegel, M. R., *Análise de Fourier.* McGraw-Hill do Brasil, Ltda., São Paulo (1976).

Stokes, S., **Glossary of Musical Terms.** Internet site, https://www.samuelst okesmusic.com/glossary.html (2019).

Story, B. H. and Titze, I. R., **Voice Simulation with a Body-Cover Model of the Vocal Folds.** *Journal of the Acoustical Society of America*, 97(2):1249–1260 (1995).

Tan, S.-L., Pfordresher, P. and Harré, R., *Psychology of Music, From Sound to Significance.* Routledge, Taylor and Francis Group, New York, USA (2018).

Tatit, L., *A Canção: Eficácia e Encanto.* Editora Atual, São Paulo, Brazil (1986).

Taylor, P., *Text-to-Speech Synthesis.* Cambridge University Press (2009).

Teixeira, J. P. R., **Modelização Paramétrica de Sinais para Aplicação em Sistemas de Conversão texto-Fala.** Dissertação de mestrado, Universidade do Porto (1995).

Thomson, S. L. and Murray, P. R., **Self-Oscillating, Multi-Layer Numerical and Aartificial Vocal Fold Models with Thin Epithelial and Loose Cover Layers.** *Proceedings of Models and Analysis of Vocal Emissions for Biomedical Applications* (2011).

Titze, I. R., *Principles of Voice Production.* Prentice Hall, New Jersey (1994).

Traube, C. and Smith, J. O., **Estimeting the Plucking Point on a Guitar String.** In *Proceedings of the COST G-6 Conference on Digital Audio (DAFX-00)*, pp. DAFX–1–DAFX–6, Verona, Italia (2000).

University, W. M., **Glossary of Musical Terms.** Internet site, Western Michigan University, https://wmich.edu/mus-gened/mus150/Glossary.pdf (2019).

Wade-Matthews, M. and Thompson, W., *The Encyclopedia of Music: Instruments of the Orchestra and the Great Composers.* Anness Publishing Ltd, London, Great Britain (2003).

Wharran, B., *Elementary Rudments of Music.* The Frederic Harris Music Co. Limited, Oakville, Canada (1969).

Wikipedia contributors, **Ut queant laxis — Wikipedia, the free encyclopedia**. [Online; accessed 21-May-2019] (2019).

Wisnik, J. M., *O Som e o Sentido: Uma Outra História da Música.* Companhia das Letras, São Paulo, Brazil (2017).

Wozencraft, J. M. and Jacobs, I. M., *Principles of Communication Engineering.* John Wiley & Sons, New York (1965).

Wyatt, K. and Schroder, C., *Pocket Music theory.* Hal Leonard Corporation, New York, USA (1998).

Wylie, C. R., *Advanced Engineering Mathematics.* McGraw-Hill Book Company, London (1966).

Yamaha, **The origins of the Marimba.** Internet site, Yamaha Group, https://www.yamaha.com/en/musical_instrument_guide/marimba/structure/structure003.html (2019).

Yudkin, J., *Understanding Music, 7th edition.* Pearson, Boton, USA (2013).

Zhang, Y., Regner, M. F. and Jiang, J. J., **Theoretical Modeling and Experimental High-Speed Imaging of Elongated Vocal Folds.** *IEEE Transactions on Biomedical Engineering.*, 58(10):2725–2731 (2011).

Zitta, S. M., *Análise Perceptivo-Auditiva e Acústica em Mulheres com Nódulos Vocais.* Centro Federal de Educação Tecnológica - CEFET-PR (2005).

Zorrilla, A. M., El-Zehiry, N., Zapirain, B. G. and Elmaghraby, A., **Pathological Vocal Folds Diagnosis Using Modified Active Contour Models.** *Information Sciences Signal Processing and their Applications.*, pp. 504–507 (2010).

Index

About the Author

Marcelo Sampaio de Alencar was born in Serrita, Brazil, in 1957. He received his bachelor degree in electrical engineering, from the Federal University of Pernambuco (UFPE), Brazil, 1980, his master degree in electrical engineering, from the Federal University of Paraiba (UFPB), Brazil, 1988 and his Ph.D. from the University of Waterloo, Department of Electrical and Computer Engineering, Canada, 1993. He has 40 years of engineering experience, and 30 years as an IEEE Member, currently as senior member. Between 1982 and 1984, he worked for the State University of Santa Catarina (UDESC). From 1984 to 2003, he worked for the Department of Electrical Engineering, Federal University of Paraiba, where he was Full Professor and supervised more than 60 graduate and several undergraduate students. From 2003 to 2017, he was Chair Professor at the Department of Electrical Engineering, Federal University of Campina Grande, Brazil. He also spent some time working for MCI-Embratel and University of Toronto, as Visiting Professor. Currently he is Visiting Chair Professor at the Department of Electrical Engineering, Federal University of Bahia.

He is founder and president of the Institute for Advanced Studies in Communications (Iecom). He has been awarded several scholarships and grants, including three scholarships and several research grants from the Brazilian National Council for Scientific and Technological Research (CNPq), two grants from the IEEE Foundation, a scholarship from the University of Waterloo, a scholarship from the Federal University of Paraiba, an achievement award for contributions to the Brazilian Telecommunications Society (SBrT), an academic award from the Medicine College of the Federal University of Campina Grande (UFCG), and an achievement award from the College of Engineering of the Federal University of Pernambuco, during its 110th year celebration. He is a laureate of the 2014 Attilio Giarola Medal.

He published over 450 engineering and scientific papers and 23 books: *Modulation Theory, Scientific Style in English*, and *Cellular Network Planning*, by River Publishers, *Spectrum Sensing Techniques and Applications, Information Theory, and Probability Theory*, by Momentum Press,

Marcelo Sampaio de Alencar

Information, Coding and Network Security (in Portuguese), by Elsevier, *Digital Television Systems*, by Cambridge, *Communication Systems*, by Springer, *Principles of Communications* (in Portuguese), by Editora Universitária da UFPB, *Set Theory, Measure and Probability, Computer Networks Engineering, Electromagnetic Waves and Antenna Theory, Probability and Stochastic Processess, Digital Cellular Telephony, Digital Telephony, Digital Television and Communication Systems* (in Portuguese), by Editora Érica Ltda, *History of Communications in Brazil, History, Technology and Legislation of Communications, Connected Sex, Scientific Diffusion, Soul Hicups* (in Portuguese), by Epgraf Gráfica e Editora. He also wrote several chapters for 11 books. His biography is included in the following publications: *Who's Who in the World* and *Who's Who in Science and Engineering*, by Marquis Who's Who, New Providence, USA.

Marcelo S. Alencar has contributed in different capacities to the following scientific journals: Editor of the *Journal of the Brazilian Telecommunication Society*; Member of the International Editorial Board of the *Journal of Communications Software and Systems* (JCOMSS), published by the Croatian Communication and Information Society (CCIS); Member of the Editorial Board of the *Journal of Networks* (JNW), published by Academy Publisher; founder and editor-in-chief of the *Journal of Communication and Information*

Systems (JCIS), special joint edition of the IEEE Communications Society (ComSoc) and SBrT. He is member of the SBrT-Brasport Editorial Board. He has been involved as a volunteer with several IEEE and SBrT activities, including being a member of the Advisory or Technical Program Committee in see veral events. He served as member of the IEEE Communications Society Sister Society Board and as liaison to Latin America Societies. He also served on the Board of Directors of IEEE's Sister Society SBrT. He is a Registered Professional Engineer. He is a columnist of the traditional Brazilian newspaper Jornal do Commercio, since April, 2000, and was vice-president external relations of SBrT. He is member of the IEEE, IEICE, in Japan, SBrT, SBMO, SBPC, ABJC and SBEB, in Brazil. He studied acoustic guitar at the Federal University of Paraiba, and keyboard and bass at the music school Musidom. He is composer and percussionist of the carnival club *Bola de Ferro*, in Recife, Brazil.